Not For Glory

A century of service by medical women
to the Australian Army and its Allies

Dr Susan J. Neuhaus
and
Dr Sharon Mascall-Dare

For Grace, Emma and Claudia,
That you may know the stories your mothers did not.

SN and SMD

This book is copyright. Apart from any fair dealing for the purpose of private study, research, criticism or review, as permitted under the Copyright Act, no part may be reproduced by any process without written permission. Enquiries should be addressed to the Publisher.

Copyright © 2021 Susan J Neuhaus

First published 2014 by Boolarong Press
Second edition published 2021 by Susan J Neuhaus

ISBN: 978-0-6451606-0-4 (paperback)
ISBN: 978-0-6451606-1-1 (e-pub)
ISBN: 978-0-6451606-2-8 (audiobook)

If you would like to submit stories of Australian women at war, both in direct military service and on the homefront during times of war, email: stories@australianwomenatwar.com.au

Front cover images
Top left-right: Captain Thi Thanh Tam Tran in Northern Iraq during Operation Habitat in 1991. *Image courtesy of the Australian War Memorial,* CANA/91/010/09.
Portrait of Lieutenant Ann Grant, a physiotherapist during World War II attached to 106 Australian General Hospital with the 9th Division. This portrait was painted by her mother, Gwendolyn Grant, on Ann Grant's return from service in Labuan, Borneo in 1946. Like many women of the Royal Australian Army Medical Corps, Lieutenant Grant did not seek public recognition for her service. *Image courtesy of the Australian War Memorial,* ART29749.
Lieutenant Georgeina Whelan in Banda Aceh in the immediate aftermath of the earthquake. *Image: Department of Defence.*
Bottom left-right: Corporal M. Patterson, Australian Army Medical Women's Service (AAMWS), works in the dispensary at 115th Australian General Hospital. *Image courtesy of the Australian War Memorial,* 100370.
Casualties arriving at Endell St Hospital, London. *Image: Wellcome Library, London.*
Colonel Susan Neuhaus, Clinical Director Role 2 NATO health facility Uruzgan, Afghanistan with child, 2009.
Back cover image
'An Operation at the Military Hospital, Endell Street' (London) was painted by Francis Dodd in 1920. The female surgical team included Dr Louisa Garrett Anderson, Dr Flora Murray and Dr Winifred Buckley. © *Imperial War Museums* (Art.IWM ART4084).

A catalogue record for this work is available from the National Library of Australia

I thought I knew about the history of women in medicine until I read Not For Glory, *the compelling stories of Australian women doctors who served our country on the battlefields of war.*

The battles faced by these remarkable pioneers were not just nation against nation. They also had to battle the dominant medical and military hierarchies of their time for professional recognition, respect and acceptance. Their inspirational stories exemplify the very best of feminist and humanist principles.

Those of us who have had the privilege of being leaders in our profession owe so much to the medical women in our history who defied convention, forming the vanguard for those of us who followed.

Professor Kerryn Phelps AM,
President, Australian Medical Association 2000-2003

More Australians at war – but women, not men; healers not killers – stories rarely heard but well told and needing to be heard.

Professor Peter Stanley,
Australian Centre for the Study of Armed Conflict and Society,
University of New South Wales, Canberra

At the outbreak of the First World War, female medical practitioners rush to volunteer in the Australian Army. They rudely were dismissed as 'too illogical or hysterical for service'. Despite this rebuttal, Australian women went on to serve with British units in theatres of war as diverse as France, Serbia and the Middle East. Their heroism and devotion to duty, and their struggle against a male dominated military hierarchy is one of the untold stories of Anzac.

In a remarkable book spanning a century of service, Susan Neuhaus and Sharon Mascall-Dare capture the lives of women too long forgotten to history. Detailed archival research, searching oral history, and vibrant participant observation shape a narrative that is bound to captivate the reader and throw new light on women's experience of war. This long overdue book is an immensely valuable contribution to the Anzac Centenary.

Professor Bruce Scates,
Chair of the Military and Cultural History Panel,
Anzac Centenary Advisory Board

Australian women doctors who tended to the wartime casualties across the globe since the Great War have been largely overlooked by historians.

The extraordinary courage, skill and resourcefulness of these women allowed them to endure the extremely trying and dangerous conditions under which they often worked, as well as the entrenched prejudice against their sex frequently shown by military authorities.

Based on extensive archival research and oral interviews, this volume brings their stories to light in a style that will engage and captivate the reader and change the way we think about the role of Australian women and war.

Professor Rae Frances,
Dean of the Faculty of Arts,
Monash University

Table of Contents

Foreword: Chief of Army   viii
Chief of Army Lieutenant General David Morrison AO

Prologue   x
Major General John Pearn AO RFD MD PhD DSc FRACP FRCP

Introduction   xiii
MEDICAL ROLE MODELS

Acknowledgements   xx
Susan Neuhaus • Sharon Mascall-Dare

ONE   001
Before Federation: women in military medicine
A WOMAN CALLED BARRY • EARLY PIONEERS • THE AMERICAN CIVIL WAR • VICTORIAN BRITAIN

TWO   012
The Great War: an empire at war
PROFESSIONAL WOMEN'S STATUS • MILITARY PLANNING • "GO ANYWAY" – A CASE STUDY IN DEFIANCE • ENDELL STREET MILITARY HOSPITAL; "DEEDS NOT WORDS" • *Captain Vera Scantlebury* • THE WESTERN FRONT • *Dr (Majeure) Helen Sexton* • THE SCOTTISH WOMEN'S HOSPITALS • ROYAUMONT • *Dr (later Captain) Elsie Dalyell* • *Miss Millicent Armstrong CdeG* • THE SERBIAN CAMPAIGN: A FORGOTTEN WAR • *Captain Agnes Bennett* • *Dr Lilian Violet Cooper* • SERVING WITH THE BRITISH • *Major Phoebe Chapple MM* • WOMEN IN THE RAMC

- *Captain Elaine Little* • *Captain Katie Ardill (-Brice) OBE DStJ*
- PHYSIOTHERAPISTS IN THE GREAT WAR • BETWEEN THE GREAT WARS

THREE 074
World War II: a corps in transition

EXCLUSION THROUGH FRAGMENTATION • WOMEN'S STATUS AS DOCTORS • *Lady (Dr) Winifred MacKenzie (née Winifred Iris Evelyn Smith)* • *Major Mary Thornton (née Kent-Hughes)* • *Major Josephine (Mabel) Mackerras* • SCIENTIFIC OFFICERS AND ALLIED HEALTH • *Lieutenant Jean Kahan*

FOUR 109
Betwixt and between: allied health care and the struggle for recognition

PHYSIOTHERAPY AND OCCUPATIONAL THERAPY • WAR REACHES AUSTRALIA: THE BOMBING OF DARWIN 1942 • *Lieutenant Joan Somerville* • *Captain Alison McArthur Campbell* • PHARMACY • *Lieutenant Gwyneth Richardson* • BRITISH COMMONWEALTH OCCUPATION FORCE – JAPAN (1946–1951) • *Lieutenant Elizabeth Wunsch* • DIETETICS • *Captain Caroline Turner* • OCCUPATIONAL THERAPY • THE VAD AND THE AAMWS: FORERUNNERS OF THE ARMY MEDIC • *Glynneath Cody (née Powell)* • VADs AND OVERSEAS SERVICE • THE HOME FRONT AND THE WOMEN'S HEALTH SERVICE • *Captain Joan Refshauge* • *Captain Joan Crosier* • *Captain Gwen Fleming* • AFTER WORLD WAR II

FIVE 151
The Vietnam conflict

AUSTRALIA AND THE WAR IN VIETNAM • *Lieutenant Dianne Skewes* • OTHER MEDICAL WOMEN IN VIETNAM • THE LEGACY OF VIETNAM

SIX 172
Other people's wars: peacekeeping near and far

CAMBODIA • *Sergeant Norma Hinchcliffe CSM* • *Corporal Elizabeth (Liz) Matthews* • WESTERN SAHARA • *Major Susan Felsche* • SOMALIA • *Corporal Kim Felmingham NSC* • RWANDA • *Captain Carol Vaughan-Evans MG* • BOUGAINVILLE • *Corporal Kerry Summerscales CSM* • EAST TIMOR • *Colonel Vikija Andersons RFD* • *Major Suzanne Le Page Langlois* • *Lieutenant Joanne Marks* • SOLOMON ISLANDS • *Captain Joanne Marks and Sergeant Kerry Summerscales CSM*

Table of Contents

SEVEN 244
A new kind of warfare: the rise of global terrorism
KURDISTAN • *Captain Thi Thanh Tam Tran CSM* • BALI • *Lieutenant Colonel Su Winter CSC* • IRAQ • *Private Vashti Henderson* • *Lieutenant Hannah Brown* • *Captain Alison Malpass* • AFGHANISTAN • *Corporal Jacqui de Gelder* • OTHER MEDICAL WOMEN IN THE MIDDLE EAST AREA OF OPERATIONS (MEAO)

EIGHT 281
Now and the future: women and medical command
Warrant Officer Class One Tracey Connors
• *Brigadier Georgeina Whelan AM CSC*

EPILOGUE 300
Reflections on service
Susan Neuhaus CSC, former Colonel RAAMC

APPENDIX 1 312
Chronology of medical women in the Australian Army

APPENDIX 2 316
Harkness Memorial Medal
C. F. Marks Award

INDEX 318

Foreword: Chief of Army

Australia has a proud and distinguished military history. From Gallipoli and the battlefields of the Western Front to the deserts of North Africa and the islands of the Pacific, to Korea, Vietnam, Iraq, Afghanistan, Timor Leste and the Solomon Islands, Australian military personnel have served our country with honour.

Yet in spite of this, our cultural identity is still predominantly tied to the men who have fought in these battles and military operations. As we commemorate the centenary of ANZAC it is important that our perspective is broadened to include the stories of all our veterans, men and women. What links us across the generations it is that war is, and remains, a fundamentally human undertaking.

This is why a book such as *Not for Glory* is so timely and important. It has been written by two talented and exceptionally well qualified women. Susan Neuhaus, CSC, has seen service as a medical officer in Cambodia, Bougainville and Afghanistan. Her co-author is award-winning journalist and now serving Army officer Sharon Mascall-Dare. They share with us the inspiring – and in some cases harrowing – stories of women serving their nation in our Army.

The combination of historical narrative and interviews with servicewomen offers a unique perspective of the service of female professionals in our Army. Together, the authors bring a fresh perspective to the history of women in the Royal Australian Army Medical Corps. As the Australian Army continues to make its commitment to equality and diversity clear, this book shows that women have always been part of our

Foreword: Chief of Army

Army's story, making important contributions in both military medicine and leadership roles. By telling the stories of a pioneering 'corps' of women, the book reflects shifts in society as well as the military. These stories are inspirational and deserve the widest circulation.

Not for Glory has done a magnificent job in reminding us that bravery, skill and compassion exist not only in the history books, but also in the current generation of female medical professionals who serve Australia here at home and wherever our soldiers deploy.

As Chief of Army, it is my privilege to introduce this book. It breathes life into an often overlooked part of our Army history and also aptly demonstrates that the Army values of Courage, Initiative, Respect and Teamwork are central components of our Army, past and present.

Lieutenant General David Morrison AO
September 2014

Prologue

It is often said that what is remembered from a scholarly lecture, or retained from reading an erudite paper, is the occasional personal story of an individual's life experiences. Such personal experiences tend to be lost in the impersonal but awful statistics of warfare; yet each tells a story which, magnified a thousand times, is the personal history of combat.

The extensive library of Australian military history continues to evolve. An important domain of research and record, hitherto relatively neglected, is the role of women in the caring corps of combat. Within the Royal Australian Army Medical Corps and, (from 1947) its predecessors, women have served in uniform as doctors, pharmacists, radiographers, physiotherapists, pathology technicians, medical logisticians and administrators. They have rendered essential and sometimes irreplaceable service both in the context of peacetime training and during hot operations. Nurses of the Royal Australian Army Nursing Corps have published a number of biographic précis as records of service. Their service is epitomised in the epitaph "Beyond All Praise", inscribed on the Nurses War Memorial on Anzac Parade, Canberra. Nevertheless, the history of these two non-combatant but frontline corps remains largely unwritten. The evolution of gender roles is an important subject for historians. The first male nurse was not commissioned in the Royal Australian Army Nursing Corps until 1972; and it was not until 1992 that Captain Tam Tran was among the first women to be decorated for exemplary military service to the nation, with the award of the Conspicuous Service Medal. The exemplary service of many doctor-soldiers, rendered far beyond the call of duty but "Not for Glory", has awaited a volume such as this.

Women doctors were denied even temporary commissions in the Royal Army Medical Corps until 7 October 1939, and permanent commissions were not instituted until 1962. In Australia, the first woman commissioned in the Australian Army Medical Corps was Lady Mackenzie (1900-1972), a titled and widowed Melbourne doctor. As the trail-blazer for female doctor-soldiers in the Australian Army, Lady Mackenzie was grudgingly commissioned in September 1940 – but only after six months' voluntary unpaid service and with the title of Honorary Captain. The War Financial Regulations in Australia were amended on 1 September 1944 to allow all prospective women officers of the Australian Army Medical Corps, irrespective of their specific professional disciplines, to be paid at the same higher rate as female officers in the other Women's Services – the Australian Army Nursing Corps, the Australian Women's Army Service and the Australian Army Medical Women's Service. Nevertheless, female doctor-soldiers received less pay than contemporary male officers of the same rank, with the same qualifications and identical postings. The 1944 Regulations constituted an important milestone in a progression towards general equality of opportunity – with subsequent appointments and promotion based on functionality and ability, rather than gender. These developments are illustrated by the individual accounts recorded in this book.

In World War I at least 17 Australian women doctors served as civilians or as doctors in the (British) Women's Auxiliary Army Corps, or were attached to volunteer auxiliary military hospitals in Europe and Africa. In this book, we read about "Major" (ungazetted) Phoebe Chapple (1879-1967) of Adelaide, who was decorated with the Military Medal for "gallantry in the field"; and Dr Elsie Jean Dalyell (1881-1948), a pathologist who was awarded a military OBE and who was twice Mentioned in Despatches, but again without a formal military commission.

The role and service of women doctors in the Defence Health Service generally, and in the Australian Army specifically, has been one of conservative evolution. It has been a path marked, like so many parallel examples, by two themes. On the one hand, gender equality was finally achieved by the altruistic service of women who proactively served as quietly-determined role models "Not for Glory", but from a personal sense of duty. A path has also been smoothed by the advocacy and the enlightened inclusive policies of a small number of senior military officers, particularly those serving in leadership roles in the Royal Australian Army Medical Corps.

The accounts of exceptional service, recounted in the pages of this book, tell a story beyond that of individual women's contributions to both peacetime and operational service. Collectively, the individual stories summarise the chronology of societal changes to women's roles in the Australian community and the successful advances to achieve not just equality in any administrative or politically-correct sense; but the rightful bestowal of dignity itself. In 1894, Professor William Balls-Headley warned against educating women to undertake professional roles in society: "If young women were allowed to undertake tertiary education, the energy needed by the uterus would be diverted to the brain …".

It took almost a century before the defence services of many western nations accepted the fact that it is not a privilege to allow women to serve equally with men in the Armed Forces, but an axiom of justice and commonsense.

This book is not primarily about gender role disadvantage in the historical sense. It is rather, in many cases, an uplifting account of service beyond self, of the service of individual courageous doctors who experienced challenges and met these beyond the normal call of duty. It is an account of proud women doing their duty individually as members, male or female, of what some call "the caring corps of combat".

Major General John Pearn AO RFD MD PhD DSc FRACP FRCP
Former Surgeon-General, Australian Defence Force
September 2014

Introduction

Time is short, and so is memory.
A.S. Walker – calling for contributions for the Official Australian Medical History
of the war of 1939-45.[1]

MEDICAL ROLE MODELS

The first Australian to be awarded a Victoria Cross was a surgeon. His name was Sir Neville Howse VC, KCB, KCMG, KStJ and he served as a soldier-surgeon in the Boer War with the New South Wales Army Medical Corps. Captain (later Sir) Neville Howse displayed conspicuous gallantry at Vredefort, South Africa, on 24 July 1900, when he ventured onto the battlefield, under intense enemy fire, to rescue a wounded soldier. Even though his horse was shot out from under him, Captain Howse continued on foot, to treat the soldier's wounds and bring him to safety. His Victoria Cross was gazetted on 4 June 1901; he remains the only Australian from a medical corps to have received such an honour.[2]

In World War II the actions of another surgeon, Sir Edward 'Weary' Dunlop AC, CMG, OBE, also became legendary. A former rugby player who had played for Australia, Weary was a 'gentle giant', standing at almost two metres tall. Captured by the Japanese, Weary cared for other prisoners of war in horrific circumstances, enduring the same privations, exhaustion and illness as his patients. Weary was tough and determined: he performed surgery and treatment with no equipment, under the most hostile of

---

[1] Walker, A. S., *Medical Journal of Australia*, 1945, vol. 2, p. 378.
[2] Bierdermann, Narelle, *Modern Military Heroes*, Random House, Australia, 2006, p. 8.

conditions.[3] He was also known for his compassionate nature: his ability to find forgiveness and make peace with his enemies also became legendary. His actions continue to epitomize the values of the Royal Australian Army Medical Corps (RAAMC) today.

A third figure, whose story is now contested, is also thought to embody those values. John Kirkpatrick Simpson landed at Anzac Cove on the first day of the Gallipoli landings, on 25 April, with the 3rd Field Ambulance, Australian Army Medical Corps. He befriended a donkey, known as 'Abdul', 'Murphy' or most commonly 'Duffy', and transported injured men up and down Shrapnel Gully to the beach. On 19 May 1915, aged 22, Simpson was shot through the heart. Although the facts surrounding his service continue to be re-examined and rewritten, the story of 'Simpson and his donkey' has entered Australian folklore. The legend (if not the facts) aligns with RAAMC values: self-sacrifice, caring and perseverance.

All three of these stories are male. Although men *and* women have contributed to the history of the Corps, prevailing narratives have privileged male voices over women's. In the chronicles of Australian Army medical history it is the men who stand out. Women's voices have been absent or silent.

This book tells the stories of many of those women. It revisits the history of the RAAMC, from its origins pre-Federation until the present day. It seeks to celebrate and commemorate the achievements of women who have challenged prejudice, exclusion and inequality to offer a significant contribution to the Corps during the past century.

In order to use their skills and training in the service of their country, many of these Australian medical women have had to fight not one, but two wars. They have contended with the powerful pressures and constraints of society; they have encountered barriers in pursuing their chosen profession. This book tells their stories sensitively, in an historical context: it marks their contribution, their courage, their legacy and their heroism. It is an attempt to address imbalance, contributing to a more comprehensive historical record.

---

[3] See: Cochrane, P. (1992). *Simpson and the Donkey: The Making of a Legend*. Burwood, Australia: Melbourne University Press.

Introduction

The story of Australian nurses deployed to conflict zones has been well researched and represented. There have been a great number of books, films and other media productions documenting the role of Australian nurses in war. At the Australian War Memorial, the exhibition entitled *Nurses: from Zululand to Afghanistan* (open from December 2011 to October 2012) showcased the service offered by these women to their country, relating the hardship, sacrifice and bravery of so many. It included records and artefacts from the Boer War until the present day, and key moments in the history of Australian nursing.

The exhibition included the story of Sister Vivian Bullwinkel, who trained as a nurse in Broken Hill before enlisting with the Australian Army Nursing Service in 1941. Posted to the 13th Australian General Hospital she served in Malaya and Singapore. After fleeing Singapore when the Japanese invaded, Sister Bullwinkel was shipwrecked and was the sole survivor of the infamous Banka Island massacre.[4] Miraculously, she managed to survive when Japanese troops machine-gunned a group of 22 nurses after they ordered them to walk into the sea. Today, the story of Sister Bullwinkel and the Banka Island massacre has become culturally significant for Australians, an affirmation of female courage when confronted by horror.

Both in Australia and overseas, modern nursing traces its ethos and foundation to the work of Florence Nightingale, who is recognised as the founder of the nursing profession. A woman who demonstrated determination and ingenuity, Florence treated casualties of the Crimean War. Known as the "Lady with the Lamp", and not the "Lady with the Stethoscope", her story has also been subject to omission and cliché, with her achievements as a statistician, administrator and social reformer not widely known.[5]

Nightingale's influence did much to make caring for soldiers wounded in war 'acceptable'. Through her prodigious reputation, nurses became 'professional' while preserving their virtues as women, protected by the strict discipline enforced by matrons. The future of the nursing profession was also shaped by her managerial and political acumen: she used her

---

[4] Shaw, Ian W 2010. *On Radji Beach*. Sydney, NSW: Pan Macmillan Australia.
[5] Lytton Strachey; *Eminent Victorians*, London (1918).

influence as a hospital administrator, her sound understanding of the nexus between politics and healthcare and her recognition of women's potential to effect social change and set an agenda for the future of nursing.

Such role models from the nursing profession are important. They continue to shape the attitudes of those men and women who follow in their footsteps, today. Our hope is that *this* book will inspire young men and women, in particular, to discover new role models of equal stature – women who have served in the Australian Army as doctors, radiographers, research scientists and medical administrators, demonstrating different skills, but the same commitment to care for those wounded by war.

Our survey of existing literature on the subject has identified omissions. In Patsy Adam-Smith's account of *Australian Women at War*, for example, less than one page is devoted to "persons and particulars"; the medical women – other than nurses – who have served their country. These women and their service have been largely forgotten. The assumption has prevailed that all medical women in the military are employed in nursing roles.

There have been attempts to address this imbalance. War is usually regarded as 'men's business', but some military historians have turned their attention to women's roles in recent years. Women's work on the home-front has been acknowledged; their labour on farms and in factories has become more widely known. Susanna De Vries's excellent biographical collection *Heroic Australian Women in War* has also made a startling contribution, with astonishing stories of women's bravery from Gallipoli to Kokoda.[6] Today, due largely to the work of De Vries and other historians, the story of Sister Vivian Bullwinkel and the 'Paradise Road' nurses have entered our recollections, adding to the fabric, and complexity, of modern Australian memory.

There are reasons why historians have chosen to focus on nurses rather than other healthcare professionals in their exploration of women's service in war. Professional roles for women in disciplines other than nursing were slower to be accepted both in civilian and military healthcare contexts. Inevitably, therefore, the contribution of nursing has been given greater emphasis; it was Australia's nurses, and those that followed, who

---

[6] De Vries S. *Heroic Australian Women in War*. 2004 Harper Collins.

Introduction

demonstrated that women not only *could* cope in conflict zones, but that their skills and service could play an important role in the overall war effort.

What is lacking, however, is a systematic and definitive review of the contribution of *other* women to Army health services from World War I to the present day. The policies that informed their service, their individual experiences and the dramatic changes that have shaped their service also need to be told.

In this text we have endeavored to unravel this complex story. It is a story of pioneering courage and determination to serve in uniform from the Federation of Australia until the present day. Today, women account for almost half of new enlistments into the RAAMC and serve in a diverse range of roles and professions. Their work, and that of their predecessors, is worthy of celebration and recognition.

This book focuses on the personal as well as the professional; it examines the careers, aspirations and struggles of individual women and the unique personal and professional challenges that they have encountered. We have sought balance through our representation of roles, geographic locations and conflicts. We acknowledge that we have not been able to recognise every woman or include every story. Those that are included highlight a particular mission or role; in all cases, we have sought to celebrate individual achievement, spirit and personality.

We have also included supplementary information about women who were not members of the RAAMC to provide context and a 'bigger picture'. This book would have lacked completeness without such inclusions. It is the women before the Corps who set the scene for later women's engagement; it is the Voluntary Aid Detachments (VADs) and the Australian Army Medical Women's Service (AAMWS) that have been the 'glue' through difficult decades, holding services (and differing working cultures) together until the Corps was mature enough to accept their skills within the professional healthcare continuum.

Sources of information for this book have come from diverse places – published literature, historical records, archival records (in Australia and overseas) and contemporary women's magazines (an amazing source of information, as we have discovered). Our work has been limited, but not hampered, by a close to complete lack of biographical records. Women change their names on marriage: in many cases records were not kept. In some instances, an individual woman's service was not considered significant enough to record. Like the women in this book who were frustrated by their attempts to pursue their vocation, we have remained persistent in seeking accurate information.

The idea of a World War I battlefield hospital run entirely by women armed with revolvers was close to unthinkable at the time, and yet it became reality. As gender roles continue to change in our society, we risk forgetting how significant these changes have been. The women in this book have opened doors that today's women of the RAAMC walk through, without necessarily noticing.

Today, the role that women play in the military remains problematic and continues to be debated. Our society grapples with issues of 'combat equality' as we deploy not just women into warzones, but wives and mothers. Distinctions between combat and non-combat roles have become less clear. Women are employed as pilots, engineers, mine-clearance experts and commanders. Suicide bombers, rocket attacks and improvised explosive devices do not discriminate by gender or by role.

The caring professions have always been seen as 'women's business', so it is not surprising, perhaps, that women have willingly taken on healthcare roles in the Army. But there are also new roles – established in just two generations – that have seen women move beyond traditional nursing roles into those of specialist surgeons, soldiers and commanders. As many medical schools around Australia graduate more females than males, there will be an increasing need to recruit medical women into the Defence Force of the future. This book intends to inspire those women by the service and adventurous spirit of their predecessors, spanning almost a century.

The focus of this book is on operational and overseas service. It is not an official history, but a compilation of narratives. The stories have been chosen to reflect the different operations in which Australian women in the RAAMC have been involved. They have also been chosen to highlight differing aspects of women's service – in terms of breaking new ground, of balancing family and service or of exceptional acts of courage and gallantry.

Introduction

We have found that these women are united by common themes that cross generations. They are women who cared enough to leave behind families and friends, sometimes to seek adventure or to escape domestic routine; sometimes out of duty and 'wanting to do their bit'. They are ordinary women who were not militaristic or out to prove a point: they were simply willing to put on the uniform of the Australian Army and use their professional skills to treat those harmed and damaged by war.

Some women received recognition through the Honours systems or from foreign governments; others did not, or received lesser awards than their male colleagues. These are women who have not actively, however, sought either glory or recognition: they simply and quietly 'got on with the job'. In these pages we hope that you will sense the quiet strength of these women. Their 'job' has been conducted under sometimes extraordinarily arduous and dangerous conditions.

We are proud of these women. Their stories are important. These Australian women have served their country with distinction and honour and demonstrated bravery, skill and compassion. In sharing their stories with you, we salute them.

Susan Neuhaus and Sharon Mascall-Dare

Acknowledgements

Susan Neuhaus

The inspiration for this book came from John Pearn, who shared with me his knowledge of the pioneering women of the RAAMC. I am indebted to his grace in allowing me to take this concept in a different direction and for his tireless support and that of Dan Kelly from Boolarong Press for their encouragement throughout its long journey into print.

So many people have helped with resources or information. These include: Annette Summers from the South Australian Military History Committee; Siobhan McHugh, author of *Mine Fields and Miniskirts*; Peter Starling from the Army Medical Services Museum, UK, for sharing documents and records related to women's World War I service; Kelly Phillips at the RACS library, who sourced rare and obscure references; Penny Russell, for access to her thesis on the early female medical graduates from Melbourne University; Cherrie Honery from DVA for her efforts and assistance with in trying to generate nominal rolls; Peter Byrne, for his wealth of knowledge about Vietnam; Philip Sharpe and Rod Cooter, for valuable information on female doctors in World War I; Charlotte Macdonald at the Victoria University of Wellington and staff at the Imperial War Museum, Wellcome Museum of Medicine, University of Melbourne Archives, Alexander Turnbull Library and the Australian War Memorial (AWM). My thanks also to Libby Stewart, for her encouragement and showing me how to navigate the AWM.

So many people have encouraged and supported me on this journey: amongst them Peter Stanley, David Horner, John Cantwell, Angus

Acknowledgements

Houston, Ashley Knoote-Parke, who first helped me realise that women's stories are important, and Donald Simpson, who is just inspirational in every way.

This work was seed-funded by a Department of Veterans Affairs 'Saluting their Service' Grant and a grant from the Australian Army History Unit, for which I am indebted. It is *prima facie* a book about Army, and would not have been possible without the support of so many of my colleagues and friends. In particular I am indebted to Michael Goodyer, who has been a rock since that first fateful exercise, and I must also acknowledge Roger Gray (who tried so hard to teach me that men and women *were* different), David Kilcullen, David Morrison, Chief of Army, who believed in my fitness as an officer (if not my 5km run times), Peter Daniel and David Saul, Padres Carl Aiken and Len Eacott and my colleagues in the Corps, particularly Jeffrey Rosenfeld, Don Beard, Richard Mallet, Nathan Klinge and Michael Clark, RSM of the RAAMC.

I also express my gratitude to my own personal support team – to Julie Learhinan and Jodie Imgraben, who keep chaos from the door – and to my family; to Mavis Evans and especially my daughters from whom I have stolen far too much time. Above all I acknowledge and thank my long-suffering husband, Peter, who has supported me, not just through this book, but has maintained the home front through each of my operational tours. Peter, your support is beyond words.

I am indebted to Sharon Mascall-Dare, whose passion and drive has seen this book become a reality. Without her involvement, persistence (and equal measures of caffeine and gin), this book would still be papers on my desk. It has been an immense journey but all the richer for your friendship and endurance.

In closing, I acknowledge each of the women included in the book. Thank you for taking us into your confidence and trusting us with your stories. You are all so inspirational and I am proud to have served in the same Corps.

There are many omissions from this list, as well as acts of grammatical heresy, for which I accept sole responsibility. There are many women whose stories are not included, including those who have served with distinction in our Navy and Air Force. I hope they will accept that this book is a tribute to *all* the men and women that have served their nation, not in pursuit of glory, but to heal those Australians wounded by war.

Sharon Mascall-Dare would like to acknowledge:

All of the women who agreed to speak honestly and generously into my microphone. I have made many new friends in researching and writing this book and enjoyed meeting every one of you. Klaus Felsche, for trusting me to respect Susan's story and his memories of her; Scot Excell, for offering his expertise; and Di and Fred Fairhead, for inviting me into their home and sharing their experiences of the war in Vietnam.

Nigel Starck, a dear friend and colleague. Nigel, you continue to inspire me with your writing and publishing endeavours: Susan and I are grateful to you for your eagle eye and editing. David Bradbury was a patient and highly professional typesetter; Dan Kelly and his team at Boolarong Press have been supportive throughout.

Army colleagues and friends who taken an interest in the project, including Doug McGuire, Dan Armstrong, Sarah Hawke, Kate Ames and the team at *Army News*. Pamela Schulz at UniSA and Matthew Ricketson at the University of Canberra continue to support my academic and long-form journalistic endeavours.

Above all, I am grateful to my family: Paul Dare, Bradley and Claudia and my parents Norman and Jean. Without your support, Paul, and your expertise in photo-editing, my contribution to this book would not have been possible. Although this book is dedicated to my daughter, Claudia, my son Bradley has become my closest friend regarding all matters military. I hope that these stories will inspire you both as they have inspired me.

Finally, I acknowledge Susan Neuhaus, who invited me to work with her on this book after we met at the University of South Australia's Narratives of War Symposium in 2011. It's been a life-changing experience, Susan. You've introduced me to many outstanding women whose stories will stay with me for life. I'll always remember our Friday night meetings at the Naval, Military and Air Force Club, discussing our latest discoveries over G&T.

CHAPTER ONE

Before Federation: women in military medicine

For I cannot think that GOD Almighty ever made them [women] so delicate, so glorious creatures; and furnished them with such charms, so agreeable and so delightful to mankind; with souls capable of the same accomplishments with men: and all, to be only Stewards of our Houses, Cooks, and Slaves.

Daniel Defoe – in his essay *The Education of Women*.[1]

Let woman be what God intended; a helpmate for a man – but with totally different duties and vocations.

Queen Victoria – in a letter to Prime Minister William Gladstone, 1870[2]

A WOMAN CALLED BARRY

In 1865 London's fashionable Cavendish Square was home to dukes, earls and viscounts. The square was known for its noble history; its symmetrical, Palladian architecture conveyed the authority of the British establishment.

On the morning of July 25, however, a secret was unveiled at Cavendish Square that would scandalise Victorian society. The story of what happened has intrigued doctors and historians for the past 150 years; only recently has the truth been confirmed. At 4 am that morning, the esteemed military surgeon Dr James Barry died in his bedroom above the square. A young

---

[1] Defoe, Daniel, 'The Education of Women', English essays from Sir Philip Sidney to Macaulay, Harvard Classics, Collier, New York, c. 1910, p. 1. Available at: www.fordham.edu/halsall/mod/1719defoe-women.asp Accessed 18 June 2013.

[2] Strachey, Lytton, *Queen Victoria: A Life*, Penguin, 1971.

Not For Glory

Dr James Barry MD (1795–1865)

servant, named Sophia Bishop, was sent to prepare his body for burial.

Dr Barry was one of the most distinguished physicians of his day. For 46 years he had served with the British Army, rising to the rank of Inspector General of Military Hospitals.[3] He had served in Jamaica, South Africa and Canada; he had introduced reforms to improve the health of soldiers and civilians alike. In Cape Town he had performed the first Caesarean section in Africa, if not the English-speaking world, saving the lives of both mother and baby.

Dr Barry was a brilliant, if belligerent character, who was also known for his quarrelsome behaviour. On one occasion he participated in a duel fought with pistols: fortunately, there was no effect to either party. On another he castigated Florence Nightingale when their paths crossed in the Crimea, leaving the founder of modern nursing bristling:

> I never had such a blackguard rating in all my life – I who have had more than any woman – than from this Barry sitting on his horse, while I was crossing the Hospital Square with only my cap

---

[3] Rae, Isobel, *The strange story of Dr James Barry: Army surgeon, Inspector-General of Hospitals, discovered on death to be a woman*, Longmans, Green, 1958.

on in the sun. He kept me standing in the midst of quite a crowd of soldiers, Commissariat, servants, camp followers, etc., etc., every one of whom behaved like a gentleman during the scolding I received while he behaved like a brute ... [Barry] was the most hardened creature I ever met.[4]

As Sophia Bishop discovered, there was more to the doctor, however, than mere brutish behaviour. Although Dr Barry had given strict instructions that his body should be left untouched in the event of his death, and wrapped in his clothes, Bishop did not comply. She lifted his nightshirt to wash his body and, reportedly, let out a scream. "The devil," she cried. "It's a woman."

The repercussions of Bishop's discovery continue to be discussed today. Until recently, the revelation remained in doubt: it seemed unlikely that Dr Barry could have concealed his gender for so long. In 2001, a surgeon from Nottingham and an archivist from Glasgow published a joint paper asserting: "Our conclusion is that Barry was not a woman. Concealment of one's sex for a year or two is probably manageable, but concealment for over 60 years, including 40 years in the British Army, is simply unbelievable."[5] Bishop's motivation was questioned: she was disgruntled after all, since she was not paid for her work on Dr Barry's body. Within a month of his death she had sold her story to an Irish newspaper.

In 2008, however, new evidence came to light. For some time it was rumoured that James Barry was the alter ego of Margaret Ann Bulkley, born in Ireland in 1789. Margaret was the daughter of Jeremiah, a grocer in Cork, and Mary-Ann, who was the sister of James Barry, a professor of painting at London's Royal Academy. In 1803, Margaret's father was sent to prison for debt, leaving his wife and daughter destitute. They contacted Professor Barry – the painter – for help and he refused, but left money for Margaret's education in his will. When he died in 1806,

---

[4] Nightingale F., Letter to Parthenope, Lady Verney (undated), London: Wellcome Institute for the History of Medicine, quoted in Kubba, A. K. and Young, M., 'The Life, Work and Gender of Dr James Barry MD (1795–1865)', *Proceedings of the Royal College of Physicians of Edinburgh*, Vol. 31, no. 4, pp. 352–356, London: 2001.

[5] Kubba and Young, *The Life, Work and Gender of Dr James Barry MD (1795–1865)*, p. 355.

mother and daughter moved to London and Margaret entered the tutelage of two family friends: a physician called Edward Fryer and a Venezuelan revolutionary – General Francisco Miranda.

Their role in Margaret's education, as part of a small group of benefactors that also included the Earl of Buchan and the Barry family's solicitor, has been previously documented.[6] In 2008, however, a retired urologist, Hercules Michael du Preez, revealed their ultimate objective. Margaret had proved herself to be a bright student. Fryer and Miranda were avid supporters of women's education. Together with Margaret's mother, the men conspired to send their protégé to university to study medicine, even though women were excluded from the profession.

Du Preez found conclusive evidence in a letter written to the Barry family's solicitor in 1809. It was signed 'James Barry' and detailed the young man's journey by ship from London to Scotland to begin his studies in medicine. "It was very usefull [sic] for Mrs Bulkley (my aunt) to have a Gentleman to take care of her on Board ship …" he wrote. On the back of the letter, the solicitor had written "Miss Bulkley, 14th December"; as a meticulous man, he wanted to ensure clarity. Bulkley and Barry were the same person.

When the scandal broke, after Barry's death, there were some who claimed to have known all along. Barry had been an enigmatic character, always immaculately dressed and slightly effeminate. Barry was constantly accompanied by a manservant who guarded his privacy zealously. There were claims that Barry had borne a child – he had engaged in a controversial relationship with the Governor-General of South Africa, Lord Charles Somerset – and Sophia Bishop claimed to have seen stretch marks on Barry's abdomen. The Army closed his files for 100 years and Barry's own physician denied the rumours. Dr McKinnon, a surgeon-major, suggested an alternative explanation:

---

[6] See: Du Preez, H. M., 'Dr James Barry: The early years revealed', *South African Medical Journal*, Vol. 98, No. 1, South Africa: 2008; and by the same author, 'Dr James Barry (1789–1865): the Edinburgh years', *Journal of the Royal College of Physicians of Edinburgh*, Vol. 42, London: 2012, pp. 258–65.

> I had been intimately acquainted with the doctor for good many years, both in London and the West Indies and I never had any suspicion that Dr Barry was a woman. I attended him during his last illness, (previously for bronchitis, and the affection [sic] for diarrhea). On one occasion after Dr Barry's death at the office of Sir Charles McGregor, there was the woman who performed the last offices for Dr Barry was waiting to speak to me. She wished to obtain some prerequisites of her employment, which the Lady who kept the lodging house in which Dr Barry died had refused to give her. Amongst other things she said that Dr Barry was a female and that I was a pretty doctor not to know this and she would not like to be attended by me. I informed her that it was none of my business whether Dr Barry was a male or a female, and that I thought that he might be neither, viz. an imperfectly developed man. She then said that she had examined the body, and was a perfect female and farther that there were marks of her having had a child when very young. I then enquired how have you formed that conclusion. The woman, pointing to the lower part of her stomach, said 'from marks here. I am a maried [sic] woman and the mother of nine children and I ought to know.
>
> The woman seems to think that she had become acquainted with a great secret and wished to be paid for keeping it. I informed her that all Dr Barry's relatives were dead, and that it was no secret of mine, and that my own impression was that Dr Barry was a hermaphrodite. But whether Dr Barry was a male, female, or hermaphrodite I do not know, nor had I any purpose in making the discovery as I could positively swear to the identity of the body as being that of a person whom I had been acquainted with as Inspector-General of Hospitals for a period of years.[7]

---

[7] Letter quoted in Kubba and Young, *The Life, Work and Gender of Dr James Barry MD (1795–1865)*, p. 355.

Dr McKinnon wrote his letter a few weeks after Dr Barry's death, and despite Du Preez's new evidence uncertainty has prevailed until the present day. While debate continues regarding Dr Barry's chromosomes – some claim that he was intersexual and of neither gender – one question remains unanswered. What is Dr Barry's legacy for the women who succeeded where he failed; for those who fought to be recognised as medical *women*?

Barry's death in 1865 coincides with the year that the first woman doctor in Britain, Elizabeth Garrett Anderson, graduated with a formal medical qualification. Perhaps Dr Anderson lived in more enlightened times (graduating half a century after Dr Barry); perhaps she was not subject to the same pressures and personal circumstances.[8] Dr Barry's achievements have largely been eclipsed by the rumours concerning his gender, but his legacy establishes one unassailable truth. If women were to succeed as military doctors in Britain, the British Empire and its dominions, they faced an uphill battle.

EARLY PIONEERS

In Europe, the Middle East and North Africa there is evidence of women physicians dating back to Ancient Greece and the dynasties of Egypt. At this point in history, men and women trained together as medical practitioners; some centres of learning continued the tradition until pre-Reformation times.

One example is the 11th century female physician Trotula, who studied at Salerno in Italy. Committed to scholarship, 'Trota' – as she is also known – became an advocate for women's health, arguing that both men and women could be infertile, a controversial notion at the time. Trotula went on to write a definitive text, comprising 27 sections, entitled: *Diseases of Women, Treatments for Women*, and *Women's Cosmetics*.[9] Some of her recommended treatments are still in use today.

---

[8] Du Preez has argued that Barry planned to abandon his disguise after he had qualified as a doctor, but circumstances forced him to maintain his disguise to pursue a career with the British Army. See *Dr James Barry (1789–1865): the Edinburgh years*, p. 259.

[9] Green, Monica H. ed. *The Trotula: a medieval compendium of women's medicine*. Philadelphia: University of Pennsylvania, 2001.

The crusades of the 11th and 12th centuries also provided active roles for women. When Pope Urban II launched the First Crusade in November 1095, his objective was simple: recapture the sacred city of Jerusalem, repel invading Turks and enforce Christian over Islamic rule. Believing participation brought redemption, tens of thousands joined the crusade. Some were armed, some were pilgrims. Many were women and children.

In the battles that followed, thousands were killed or wounded. The fortunate made it back to Constantinople, the capital of the Byzantine Empire. In the centre of the city there was a hospital founded by the Emperor, Alexios I Comnenus. With 10,000 beds, it was managed and administered by his daughter, a teenager at the time. Her name was Anna Comnena.

Historically, Anna has been held up as a pioneer in healthcare provision, recognised by the nursing profession. Given her royal status (she was born 'in the purple'), it is unlikely that she was personally involved in patient care. Anna's specialty was medical administration and command. Aged just 12 when Pope Urban II declared 'Holy War', Anna spent her teenage years mastering the structure and organisation of a major medical facility. She also became a keen observer of the clinical condition: she was the first medical practitioner to write about psychosomatic disease.

The experiences of Anna and Trotula were exceptions, however. In Europe, women's access to medical training and professional practice remained limited from the Dark Ages until the 20th century. This was in contrast with other parts of the world: Dr Barry's sponsor, the Venezuelan revolutionary General Francesco Miranda, had his protégé earmarked for a post in South America at one point, where women doctors were not unusual. In Australian Indigenous culture, the presence of women as healers, bone setters and obstetricians is also long-standing and continues to be conveyed through traditional law, sacred sites and rituals.

In Britain, the dissolution of the monasteries in the 16th century saw the dissolution of the nunneries, previously houses of female learning, skills and healing. From 1614, no woman could obtain a licence to practise surgery in Britain; in 1617, they were excluded from working as apothecaries and pharmacists.[10] In 1642 it also became increasingly difficult for British women to obtain a midwife's licence. In order to practise, they were required to complete an examination presided over by six surgeons and six midwives – all of whom were male.[11]

THE AMERICAN CIVIL WAR (1861–65)

Discrimination against women doctors also prevailed in Britain's dominions and colonies, even after they rebelled against British rule and established independence.

The American Civil War remains the bloodiest and most divisive conflict in the history of the United States: the death toll for the Battle of Gettysburg alone was 43,000. As the country was torn apart, both sides recognised the need for battlefield doctors. Still, women encountered discrimination.

By the time war broke out in 1861 only a handful of women had graduated as doctors in America. Despite their qualifications, they found it impossible to convince military authorities to commission them. Among them were Dr Elizabeth Blackwell, America's first female medical graduate; Dr Ester Hill Hawks, who set up a hospital for wounded black soldiers; and Dr Sarah Clapp, who served as a surgeon.[12]

Their most notable contemporary was Dr Mary Edwards Walker. She was the only woman surgeon to receive an official appointment during the American Civil War and the only woman ever to receive the Congressional Medal of Honour, the highest military award in the United States.[13]

At the age of 23, Mary graduated from Syracuse Medical College in New York. Recently divorced at the outbreak of the war, she was determined to join the Union Army as a surgeon. Despite unsuccessful applications to

---

[10] See: W. Bonser, *The Medical Background of Anglo-Saxon England*, Wellcome History of Medicine Library, London 1963.

[11] Hutton Neve, Marjorie, *This Mad Folly*, Library of Australian History, Sydney: 1980, p. 16.

[12] Ester Hill Hawks, *A Woman Doctor's Civil War*, Gerald Schwartz, ed., Columbia: University of South Carolina Press, 1984.

[13] Favor Lesli J. *Women Doctors and nurses of the Civil War*, Rosen Publishing Group, Inc. New York: 2009.

the Surgeon General and a personal appeal to President Lincoln she was unable to secure a commission.

Undeterred, she undertook voluntary work at first, training under the well-known nurse Dorothea Dix. Mary then invented herself as an officer, dressing in a blue uniform. Her attire was both stylish and shocking for its time: she wore gold-stripped trousers, a felt hat with gold braid and a green surgeon's sash.

Through the following year Mary worked unflinchingly as a battlefield surgeon, witnessing unimaginable horrors. Field hospitals of the day were ramshackle, with hastily acquired accommodation and straw-filled mattresses. This was a time before 'germ theory', when doctors came to understand bacteria and antisepsis in the late 19th century. Patients were accommodated where they could, amid the detritus of human waste, amputated limbs, camp dogs and garbage.

During the winter in early 1864, Mary made frequent forays behind enemy lines to carry medical supplies to desperate civilians caught up in the fighting. During one of these missions – dressed in full Union uniform – she was captured by Confederate troops, accused of spying and became a prisoner of war.

For the next four months she was imprisoned and derided as a curious debasement of womanhood: "A thing that nothing but the … depraved Yankee nation could produce – a female doctor."[14] During this time as a prisoner of war, her vision was permanently damaged.

Mary was proud that, when she was exchanged along with other Union prisoners, her 'price' was commensurate with a male surgeon. She was also paid the sum of $432.36 in back salary for her services between March and August 1864. For her, this affirmed her legitimate position in the military.

Dr Walker was awarded the Medal of Honor for her services. This unique honour for those who, 'distinguished themselves by their gallantry in action, and other soldier like qualities', has not been awarded to another woman since. In 1917, the requirements of the award were redefined to

---

[14] Karen Zeinert, *Those Courageous Women of the Civil War*, Brookfield, Milbrook Press: 1998, quoting General Joseph E Johnston of the Confederate Army.

exclude non-combat roles and Mary's was rescinded, along with 910 others. Regardless, she wore her medal with pride for the remainder of her life. In 1977 the honour was restored to her, posthumously, by President Jimmy Carter.[15] To this day she remains the only woman to have received it.

It is significant that American society remained scandalised throughout the Civil War period; that women should see men in a state of undress, far less participate in their medical treatment. Still, the work of Dr Walker and her colleagues encouraged a generation of women to pursue professional medical training, and in the years to come they continued to apply their skills on the battlefield.

Victorian Britain

Back in Britain, the accession of a female monarch – Queen Victoria – began to shift attitudes towards women's education and employment.

The Queen, however, was openly opposed to women doctors and expressed her displeasure at what she saw as a corruption of morality. In May 1870 she wrote to the British Prime Minister, William Gladstone. She condemned the:

> … mad and demoralizing movement of the proposal to place women in the same position as to professions – as **men** – and amongst others, in the **Medical Line** … She is not only most anxious that it should be known that she not only disapproves but **abhors** the attempts to destroy all propriety and womanly feelings which would inevitably be the result of what has been proposed… but to tear away the barriers which surround a woman … Would be to introduce a total disregard of that which must be considered as belonging to the rules of the principles of morality…

She went on to state:

> Woman will become the most hateful, heartless and disgusting of human beings were she allowed to unsex herself … Let woman

---

[15] See: Mary Edwards Walker. 2014. The Biography.com website. Available from: http://www.biography.com/people/mary-walker-9522110 Accessed 14 Jun 2014.

Before Federation: women in military medicine

be what God intended; a helpmate for a man – but with totally different duties and vocations.[16]

It is ironic that the Queen herself did not adhere to these characteristics. She ruled an Empire spanning three-quarters of the globe; her beloved husband, Albert, served as a consort rather than King; and, when she was born, the infant Victoria was delivered by a woman doctor, Fraulein Siebold, who was brought from Germany to England to deliver the heir-presumptive.[17]

Despite her personal opposition, and her determination to 'rescue from immorality young women […] … who contemplated a medical career', her position and leadership, as a formidable female monarch, were indicative of change.

Her death in 1901 came just three weeks after Australia's inauguration as an independent nation. In 'the Colonies', Australian girls from the middle and upper classes had been enjoying considerably more freedom than their English counterparts for some time. The Sydney Medical School opened its doors to women in 1885, followed closely by Melbourne University in 1887.

Still, these female pioneers of Australian medicine continued to encounter obstacles and disapproval. They were usually from privileged backgrounds and of 'independent means'; they funded their studies and professional expenses privately. A commitment to social and community service was pervasive; they felt a duty to contribute to society, even though society was hesitant – and obstructive – in response. By the time of the Boer War 1899–1902, Australian women were already doctors in their own right, but public opinion was such that no role in military life could be contemplated. It was not until World War I that women would demonstrate ability in military medicine.

When war broke out in 1914, and the threat of war descended across the British Empire, Australia's medical women showed no reticence in rallying to a distant bugle.

---

[16] Strachey, L. *Eminent Victorians* 2006 Penguin Books London, England pp. 111–161

[17] Hutton N, p. 21.

CHAPTER TWO

The Great War: an empire at war

These quiet women ... were the true pioneers. They did not call upon the world to listen to what women might, could or should do under quite different conditions; they simply did – under existing conditions – first the thing that needed to be done, then and there.

Ruth Bowden OBE (1915–2001) – Professor of Anatomy, Royal Free Hospital, and advocate for women's education.[1]

If young women were allowed to undertake tertiary education the energy needed by the uterus would be diverted to the brain, rendering them infertile.

William Balls-Headley (1841–1918) – gynaecologist and Professor of Obstetrics, Melbourne University.[2]

On 28 June 1914 Archduke Ferdinand of Austria was shot by a Serbian student in Sarajevo. The assassination triggered a cataclysm of bloodshed across Europe. In response to Britain's declaration of war on Germany, Australia's Prime Minister, Andrew Fisher, uttered the words that would profoundly affect so many Australian lives.

Australians will stand beside our own to help and defend Britain to our last man and our last shilling.[3]

---

[1] Bowden, Ruth E., in the foreword to Crofton, Eileen, *The Women of Royaumont – A Scottish Women's Hospital on the Western Front*, Tuckwell Press, 1997.

[2] See, Balls-Headley, W., *The Evolution of the Diseases of Women*, Smith, Elder, London, 1894.

[3] Murphy, D. J., 'Fisher, Andrew (1862–1928)', *Australian Dictionary of Biography*, National

Implicit in the Prime Minister's words, and evident in the response of the Australian Army at the time, was that women need not apply.

Professional women's status

Towards the end of the 19th century, professional prospects for women were changing. By 1914 nursing was regarded as a socially acceptable female occupation: it was compatible with their preparation for future roles as wives and mothers; it was also acceptable for women from higher social classes. The standards enforced by Florence Nightingale had helped to change attitudes towards nurses: they were now regarded as women above reproach, subject to the strict discipline of a matron.

By contrast, female doctors were regarded as an enigma and treated with suspicion. They remained an affront to the Victorian ideal of womanhood – why could they simply not accept their place? In the early years of the 20th century, intelligence and independence were not considered to be female virtues. The thought that women would join with men in anatomical dissection laboratories, operating theatres or the examination of (undressed) male patients remained rather shocking.

By 1914, Australian women had won their struggle for access to medical education. Despite a number of setbacks and refusals, seven young women were admitted to the Medical Faculty of Melbourne University in 1887, some 12 years after Elisabeth Garrett Anderson obtained her licence to practise medicine in England. The first to graduate in Australia were Clara Stone and Margaret Whyte in 1891, followed by Helen Sexton and Elizabeth O'Hara in 1892. Dr Stone recalled:

> When we entered medicine we knew we were definitely not wanted. The staff were always most courteous ... but the boys didn't want us there – as their attitude showed clearly ... When Margaret distinguished herself academically by winning not only several prizes but two coveted scholarships the men were absolutely furious, their attitude in general was that if we

---

Centre of Biography, Australian National University, http://adb.anu.edu.au/biography/fisher-andrew-378/text10613, Accessed 1st March 2013.

girls hadn't forced our way in the men would have won the scholarships and some of them never forgave either of us."[4]

In Australia, it was not considered feasible to open a separate medical school so that women could study dissection and anatomy independently of their male colleagues. As a result, lessons were held together, but access to the anatomy laboratories and dissecting rooms was carefully timed so that male and female students were separated, to preserve the modesty of all.

After graduation there were further barriers to women seeking suitable hospital posts. It was one thing to allow women to study and graduate in medicine; it was another to allow women to work alongside men in the profession. The experiences of Dr Susie O'Reilly, are a case in point. She graduated from Sydney University in 1904 and was refused appointment to a hospital post. Despite passing fourth on the Merit list in her final examinations (normally an assurance of a position), the Board of Directors of the Sydney Hospital refused to appoint her on the grounds of her gender:

> It would not be compatible with the best interests of the hospital to appoint a lady doctor … and at the present time at any rate, there [is] not sufficient accommodation in the hospital for a resident lady doctor.

A vitriolic and public debate followed fuelled in part by support from Sydney's *Daily Telegraph*. Ultimately, Dr O'Reilly did join the hospital staff, but she was obliged to reside in the nurses' quarters.[5]

At the outbreak of the Great War, therefore, women doctors were neither universally accepted nor welcomed in the profession. The attitudes of the medical establishment at the time were summed up in the comments of Sir Thomas Anderson Stuart (1856–1920), Professor of Physiology and Dean of the Faculty of Medicine at Sydney University:

> The proper place for women is in the home, and the proper function for a woman is to be a man's wife, and for women to

---

[4] Hutton Neve, *This Mad Folly*, Sydney: Library of Australian History, 1980 p 30.
[5] ibid. p. 79.

be mothers of our future generations ... I have come to the conclusion that, within certain limits they (women doctors) have played a useful part in medial life, but there are limits and they will never in my opinion take the place of or be equal to men in general medical work.

His advice to women seeking a place in the medical profession was that "they would be much better employed if they got a nice frock and a nice man".[6]

When Prime Minister Fisher made his historic commitment to stand by Britain in 1914, it was not expected that women would serve their country or the Empire. The primary role of women doctors would be civilian work or to fill positions left vacant by male doctors who took up commissions in the army.

MILITARY PLANNING

When war broke out, the Australian Army based its response on Britain's military structures of command and control. This also applied to its medical teams. From the outset, Australia contributed medical teams and personnel to support plans devised by Britain, representing the Empire as a whole. Confusion still reined: there was considerable debate about how many doctors Australia should send, for example.

Ceding significant medical control to Britain was controversial. Although Australia accepted that it would be impractical to have Australian men attended by Australian doctors, there were concerns about the care afforded to Australian troops in British hospitals; concerns remediated by a policy of repatriating all soldiers deemed unlikely to recover within six months.[7]

Despite such concerns there was a significant mobilisation of Australian medical professionals determined to 'do their bit' and serve as doctors alongside the thousands of young men who signed up. The best-known

---

[6] See ibid, p. 54, citing Epps, William, *Anderson Stuart MD*, Angus and Robertson, 1922.

[7] Featherston, R.H.J., 'On Australian Army Medical Services Overseas', *Australian Parliamentary Papers 1917–19*, IV, No. 132, pp. 18–19, p. 32 and pp. 36–7.

group of Australian doctors, who went to England to join Britain's Royal Army Medical Corps (RAMC) in 1915, were known as 'Kitchener's One Hundred'. These Australian men distinguished themselves on the shores of Gallipoli and in the battlefield hospitals of the Western Front and Egypt, demonstrating heroism and professionalism in pursuit of their work as doctors.[8]

In terms of numbers, there was considerable pressure on British medical units throughout the war. Still, the Director-General of the Australian Army Medical Services (AAMC), R.H.J. Featherston, stated unambiguously in 1917 that he had more doctors than he could use. He estimated that 994 Australian medical officers had enlisted and were serving abroad with the AAMC, with a further 1,362 on active and reserve lists. In addition, there were between 250–300 civilian medical practitioners providing part-time services.

In 1915 there were 129 women listed in *Butterworth's Medical Directory* – a reliable gauge of Australia's capability and capacity in the field of medicine at the time. Although the overall numbers of Australian doctors who enlisted officially remains unclear, it is estimated that there were approximately 3,000 who served in one capacity or another: none of those listed were women.

When war broke out, caring for the wounded in battle was still considered a task for men: women were not to be subjected to the horrors of warfare. It was also considered impractical to have women associated with an army on the march, in case they were killed or taken prisoner. Accepting female nurses posed enough of a challenge, never mind considering women in any other roles. In his book *Our War Nurses*, Rupert Goodman notes:

> Opposition to women serving in the Army came in part from those who thought women should not be exposed to the horrors of war, but the strongest opposition came from the army itself, from senior officers who were of the view that the army was a man's world, that there was no place for women in the structure itself. There was no objection to those volunteers outside of the army, ladies of the Red Cross and VADs, who could help in a

---

[8] Likeman, Robert, *Gallipoli Doctors*, Slouch Hat Publications 2010.

small way in base hospitals. But the idea of enlisting them in the army, giving them rank, sending them overseas, incorporating them into the Army Medical Corps was an anathema to some. This was to be a testing ground for women, who in civilian life were making tentative steps to move into new occupations. Army nurses were to be the frontrunners in demonstrating that they had a place in a man's world.[9]

Women at the time were considered 'too illogical or hysterical' to serve as medical officers or surgeons; they would be a distraction and cause more difficulty than they were worth. Similarly, in Britain, the War Office refused to entertain the idea of medical women serving in military hospitals: those who wished to serve offered their services to Allied governments elsewhere in Europe. French, Belgian and Balkan forces were suffering severe shortages of manpower, particularly in medical support. These governments were grateful for assistance, even if it did arrive wearing a skirt.

In America too, women doctors were not accepted within the Army Medical Corps at this time, although they were allowed to serve as contract surgeons. This meant that were allowed to fill in when the U.S. Army was unable to find a suitable male physician. They were employed with no rank, no standing, no opportunity for promotion and no pension or disability provisions.

These barriers to women's employment as army doctors did not come down, solely, to concerns about competence. Equally relevant were concerns about their suitability and the reactions of men. As one American colonel on the staff of the U.S. Surgeon General stated:

> Such a position, in my judgment, is not befitting a woman. There are obvious reason why it is not desirable that they should be called upon to examine large numbers of men stripped to the skin … and come into contact with large numbers of men drawn from all classes of society, many of whom would not understand the precise position of the woman and think of her only as a woman.

---

[9] Goodman R., *Our War Nurses: The History of the Royal Australian Army Nursing Corps 1902–1988,* Boolarong Publications, Brisbane, 1988, p. 18.

> Furthermore there are few women who are physically qualified to endure the fatigues and vicissitudes of a campaign.[10]

Such concerns were not extended to nurses, who were equally exposed to large numbers of men 'stripped to the skin' and 'drawn from all classes of society'. Indeed, in 1917, the British Army replaced some of its field doctors with nurses, who had been trained by Americans, to provide anesthesia. Reassuringly, perhaps, for their male superiors, these nurses retained their segregated status in the military medical hierarchy and remained under male authority.

Similarly, Winston Churchill, Britain's First Lord of the Admiralty, was also sceptical about women's ability to lead teams of army medical personnel:

> The command of Medical Field Units involves leadership and discipline, and at times very great strain and hardship to which women would only be equal in very rare cases.[11]

Women doctors were not allowed, therefore, to join the Australian Army or serve as officers. In fact, when one of them (Dr Agnes Bennett) did attempt to enlist, she was told to "go home and knit". They were welcome to volunteer their services to organisations such as the Red Cross, but those who wanted to treat *soldiers* or practise battlefield medicine had to find another way.

Consequently, when Australian troopships set sail on 1 November 1914 to establish the first Australian General Hospitals for the First Expeditionary Force, the only women on board were the 25 nurses of the Australian Army Nursing Service or AANS.

"Go anyway": a case study in defiance

After their long struggle to obtain a medical education, some women were anxious to serve in a military capacity. They had not worked so hard, only

---

[10] Bellefaire J. and Graf M., *Women Doctors in War*, Texas, A&M University Press, 2009, p. 34.
[11] The War Office in document dated 2 May 1919 in CMAC:SA/MWF/C.163 Women at War 174.

to be denied the opportunity to serve their country. They were determined to test their newfound skills in the harshest of learning environments – on the battlefield itself.

Australians were informed of news from the front through newspaper reports. Every month the recently established *Medical Journal of Australia* listed the names of doctors joining the Australian Army Medical Corps. From 1915 the *MJA* published numerous journal articles informed by Australia's wartime experiences, detailing cases of war wounds, the management of tuberculosis and conflict-related medicine. Advertisements for Australian male doctors to join the war effort were included in every edition, but there was no suggestion that women could apply.

A number of Australia's medical women saw no reason why their skills could not be employed alongside men. Highly educated and often affluent members of Australian society, they were women who cared little for the opinion of their male colleagues – some had endured years of hardship and even ridicule during their training.

Keen to demonstrate their mettle, their priority was the welfare of soldiers and civilians. They saw service at or near the front as not only their duty, but a genuine opportunity to prove themselves. Also, as was the case for many young Australian men at the time, they saw the opportunity to serve abroad as the adventure of a lifetime. These were women who remained undeterred by the reluctance of the Australian Army to use their services. Accustomed to rejection, those who tried to join the Army in 1915 were neither surprised nor discouraged when the answer was 'no'. They simply found another way.

In total, 15 of the 129 female medical practitioners registered in Australia at the time overcame barriers and bureaucracy to serve as doctors during the Great War, working in France, Serbia and Malta. They served as doctors, surgeons and pathologists; they were employed by a number of different organisations, including the British War Office, the RAMC and the Scottish Women's Hospitals.[12]

Many were given the same responsibilities as male doctors, and were

---

[12] Mitchell A. M., *Medical women and the medical services of the First World War*, Sydney, published by the author, 1978 p. 20.

subject to the same rigors of service, risks and dangers. This was despite claims by the British War Office that medical women could only offer 'limited service' in the field.[13] Other medical women worked as masseuses (physiotherapists of their time) and ambulance drivers. They treated patients in hospitals in England, on the Western Front and Egypt; they also worked in makeshift field units during the Balkans campaign in Greece.

Despite assertions by Surgeon General Featherston that there were 'more than enough' Australian doctors, British medical units were under increasing pressure throughout the early stages of the war, due to a lack of manpower. Relieved later by the arrival of medical teams from the United States, by the end of 1916 the need for more staff was paramount and Britain called for female doctors to enlist.

Women were then permitted to join War Office sanctioned military hospitals in Britain and overseas. They were, however, denied official RAMC rank and the associated authority, status, uniform and badging. These anomalies (discussed below) affected not just their service, but the recognition of that service subsequently, through the award (or non-award) of medals.[14]

Despite the change of policy in Britain allowing women to enlist, the Australian Army did not admit female doctors until World War II. Those Australian women who did serve in the Great War as doctors were required to do so under the flag of either the Imperial British Forces, or one of their Allies. Some joined British Army units, serving in France, Malta, Egypt or Salonika with the RAMC. One of the first Australian women to do so was Dr Agnes Bennett, who joined the RAMC in Egypt en route to Britain. Other doctors such as Phoebe Chapple served as part of Queen Mary's Auxiliary Army Corps both in England and in France. Doctors Rachel Champion, Emma Buckley, Eleanor Bourne, Vera Scantlebury and Elizabeth Hamilton-Browne[15] were employed in the War Office-funded

---

[13] Leneman, Leah, 'Medical Women at War 1914–1918', *Medical History*, no. 38, 1994, p. 161.

[14] Watson, Janet K., *Fighting a Different War: Experience, Memory and the First World War in Britain*, Cambridge University Press, 2004, p. 68.

[15] The experiences of Vera Scantlebury are explored in more detail later in this chapter. Rachel Champion, a medical practitioner from Melbourne, served as an assistant surgeon at Endell Street Hospital. Emma Buckley, from Sydney, also served at Endell Street. with the Women's

Endell Street Military Hospital in London. Some served in multiple theatres or with more than one Allied agency.

Like the men who served alongside them, they volunteered for a 'grand adventure', unaware of the realities they would encounter. Differing in age and experience, they were united by a determination to support their nation and her Allies at war. In the months and years to follow, their service changed them: it shaped and defined their personal and professional lives.

ENDELL STREET MILITARY HOSPITAL: "DEEDS NOT WORDS"

London's Endell Street Military Hospital was the brainchild of Dr Louisa Garret Anderson, the daughter of the first female doctor in Britain (Elizabeth Garrett Anderson). Anderson established the facility with her friend and colleague, Dr Flora Murray.

Both were well aware of resistance to their involvement in 'war work' as women doctors. Still, they were determined to carve out roles in military medicine. Like many women doctors of the time, they had been educated in an environment of militant feminism and activist politics. The decision by feminist groups to suspend their more radical, suffragette activities during the war meant that a number of highly educated women were seeking a cause worthy of their energy and support. Paradoxically, the war provided an opportunity for these women to involve themselves in 'men's games', gaining respect by saving, rather than taking, lives.[16]

For Dr Anderson and Dr Murray, a first opportunity arose in September 1914. Taking advantage of the administrative shambles that prevailed at the time, they managed to bypass English officialdom and, with the enthusiastic agreement of the French embassy, set up a hospital for wounded

---

Hospital Corps. In January 1916, Eleanor Bourne went to England at her own expense and served as a lieutenant with the Royal Army Medical Corps at Endell Street. Promoted to major in 1917, she also became a medical officer to the Queen Mary's Army auxiliary corps. Elizabeth Hamilton-Browne, a medical practitioner from Sydney worked at Endell Street before serving with No. 19 General Hospital in Egypt. She later became a Medical Officer in France. For further details see https://sites.google.com/site/archoevidence/home/ww1womendoctors

[16] Mitchell A. M, p. 10.

soldiers in Paris. Based in the Hôtel Claridge, the hospital was staffed and equipped entirely by women. Uniforms for doctors and orderlies were designed and manufactured in Britain and they were transported, along with staff and equipment – including an X-ray machine for locating bullets and operating theatre equipment – from Victoria Station to Paris on 15 September 1914. Within four weeks a fully functioning hospital had been hastily constructed. It was largely funded by private donations and supplied with expensive bed linen obtained from the stores of the Hôtel Claridge put to patriotic use. [17]

A feat of organisation, the hospital accepted patients almost immediately, receiving casualties directly from the front. Most of them were victims of tetanus, gangrene, shock and sepsis. Despite being technically a French hospital, the patients included both French and British casualties.

Early battlefield evacuation plans were clearly inadequate. Many of the patients laid on stretchers for days at a time, with their wounds not dressed or changed since their arrival. The condition of the men was described by Viscount Esher during a visit to northern France:

> A train came in, and these shattered men were lifted into cattle-trucks and placed on straw. The condition of these trucks was indescribable.[18]

Dr Garrett Anderson wrote of her first few days:

> Sometimes I am in the theatre from 2 to 9 or 10 at night and have eight or more cases ...
>
> The cases come to us very septic and the wounds are terrible. Today we are having an amputn [sic] of this, 2 head cases perhaps trephine and five smaller ones ... the cases are very <u>heavy</u>. [sic] Especially the severe fracture of thighs. They need 4 people at least to dress them.[19]

---

[17] Geddes, J. F.

[18] Lord Esher's War Journals, Memo for Lord K: The War Office and the Red Cross (undated, end of September), ESHR 2/12 Churchill Archive Centre.

[19] Letter from Louisa Garrett Anderson to her mother quoted in Geddes, Jennian F., *"Women as army surgeons": the Women's Hospital Corps*, MA thesis, London Metropolitan University, 2005

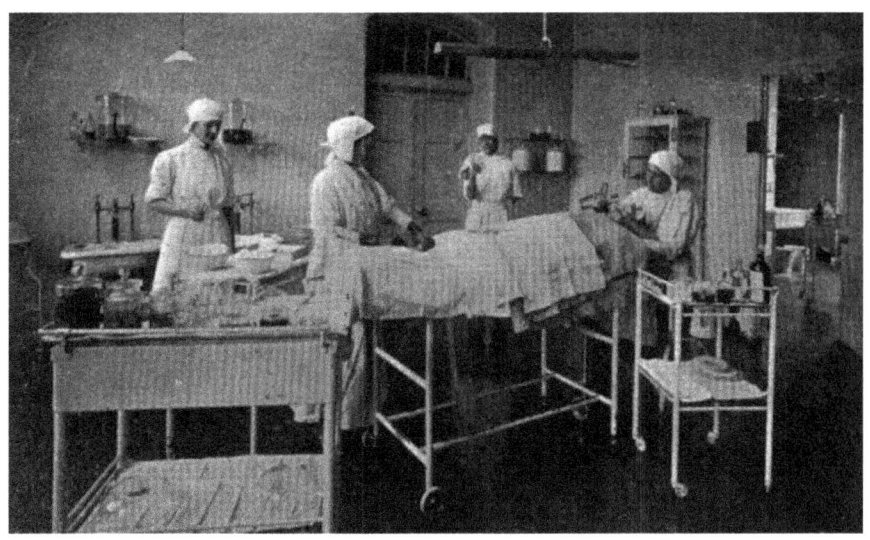

Operating Theatre, Endell St Hospital, London.
Image courtesy of the Wellcome Library, London

Over the next few months Dr Anderson and Dr Murray proved that they were not only up to the job, but that they possessed the discipline and stamina required for war surgery and the capacity to manage and administer a hospital. This would prove to be pivotal in supporting their case for a similar hospital in Britain.

As the front moved away from Paris, the Hôtel Claridge hospital (also known as the Hospital Anglo-Belge) was closed for business, as was a similar hospital at Wimereux, near Boulogne, by the shore of the English Channel. Wimereux had also been staffed entirely by women. Returning to London, and emboldened by their success in France, Anderson and Murray began to lobby the War Office for a chance to work with the RAMC and establish a military hospital in London.

The Endell Street Military Hospital opened in May 1915. Partly in response to need, and partly as a way of dealing with the 'women problem', the War Office approved the establishment of a female-run military hospital in an old workhouse in Bloomsbury. It was a short walk from the British Museum and close to the main railway stations, receiving convoys

---

(held at the Women's Library, London), p. 27.

of wounded by ambulance trains arriving directly from France. London was far enough away from the fighting or danger to be a 'suitable' place for women. Anderson and Murray received official sanction to open their 'women's hospital' and were appointed as Commanding Officers.

Flora Murray, four years older than Dr Anderson, was the "Doctor-in-charge". Promoted to the equivalent of Lieutenant-Colonel, she was the only woman recognised at that level by the British Army.[20] Dr Anderson, who was aged 41 when war broke out, was Major equivalent, while the rest of the senior medical staff were appointed as Captains.

Within a week, all of the hospital's 500 beds were full with casualties arriving from heavy fighting in France. The hospital remained operational until 1919 and some 26,000 patients – mostly British but also Australian, Canadian or members of British Dominion troops – passed through the wards.[21] For the duration of the war, all members of staff (with the exception of a small unit of RAMC orderlies and their sergeant-major) were women. It was a hospital where female doctors treated thousands of male patients – unprecedented on such a large scale.

Endell Street, as it was commonly known, was run as a professional military hospital. Although the women did not have formal commissions, they did have rank and drew their pay and allowances accordingly. Although not technically entitled to use their rank or wear the King's uniform, the doctors adopted military-style khaki uniform suits with ties and RAMC lapel badges and buttons. They referred to each other by surname and proudly defended the status of their hospital as a female-run institution officially accredited by the War Office.[22] The fact that women were not given commissions was to remain a contested issue, but the War Office clearly regarded the women as 'attached on short-term contracts' rather than being a part of the RAMC.

Their salaries were drawn under Royal Warrant from army funds. The women were not formally 'in' the army, however, and had no entitlement to service rates of income tax or veteran benefits. Unlike their male

---

[20] Geddes, 'The Women's Hospital Corps: forgotten surgeons of the First World War'.
[21] Leneman, 'Medical Women at War'.
[22] Geddes.

counterparts, they travelled third class on trains. Such inconsistency – and as some women observed at the time, injustice – prevailed throughout the Great War. Although pragmatism won out in some cases, and women in the medical services received greater respect from their male peers, bureaucracy continued to treat them differently. The distinction between men's and women's service went on to be reflected in the official histories of the war: it explains, in part, the lack of recognition that these women have received.

Captain Vera Scantlebury

One of the five Australian female doctors who worked at Endell Street was Vera Scantlebury, later Vera Scantlebury-Brown. A graduate of Melbourne University, she paid her own way to England and the war.

Her recollections of her time at Endell Street are recorded in 19 volumes of diary letters written between March 1917 and February 1919.[23] Preserved in the University of Melbourne Archives, her prolific letters provide insight into her conditions of service and the leadership of the women she worked for. They also record her own personal journey from a somewhat naïve and inexperienced doctor to a mature surgeon and clinician.

Vera was a diminutive figure, aged 28 and nicknamed 'the little lieutenant doctor' by her patients at Endell Street. Her youth and inexperience were immediately recognised by Doctors Anderson and Murray. She initially found their mentoring and close supervision frustrating having, in her own mind, already established her medical credentials as a resident doctor at the Melbourne Children's Hospital. Her induction was not unusual: Anderson and Murray took time to mentor and supervise all new members of staff individually, quietly assessing their capabilities and characters.[24]

---

[23] Scantlebury, Vera., *The Scantlebury-Brown Papers 1917–1918*, held in the University of Melbourne Archives.

[24] Sheard, H., "They will both go to heaven and have crowns and golden harps": Dr Vera Scantlebury Brown and Female Leadership in a First World War Military Hospital', *Founders, Firsts and Feminists: Women Leaders in Twentieth-century Australia*, eScholarship Research Centre, the University of Melbourne, 2011. Available at: http://www.womenaustralia.info/

Captain Vera Scantlebury with her brother Cliff, during a visit to London.

Despite her initial self-confidence, Vera struggled with emotional turmoil during her first eight weeks in the operating theatre. As a result, Dr Anderson – who was well aware of the shock and fatigue caused by first exposure to the trauma of war wounds – insisted she take a week's leave. There is no doubt that Vera found her first six months at Endell Street personally and professionally challenging. Her writing illustrates her angst in coping with a military structure, her separation from her fiancé in Australia and the professional challenges of war surgery.

> At present my mind is a confusion of military etiquette and rules, unusual methods for arrangements for operating etc – a fury and turmoil in my brain against this dreadful war causing this inexorable suffering.
>
> I am not at all keen on military surgery but suppose I will get used to it and do it better than at present but I think it is horrible.[25]

---

leaders/fff/pdfs/brown.pdf Accessed 29 November 2012.
[25] Sourced from *The Scantlebury-Brown Papers*, 21 June 1917, Vol. A3, p. 60; 2 May 1917, Vol. AD5, p. 22; 25 Jun 1917, Vol. A3 p. 68; and 24 October 1918, Vol. A14, p. 6.

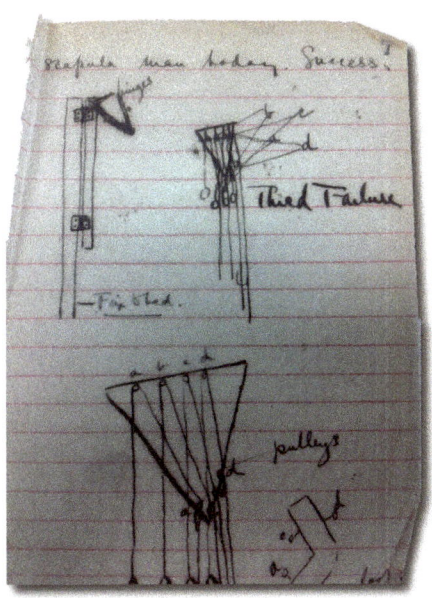

Captain Vera Scantlebury's diary entry (undated) describing her attempts at scapula traction for a casualty with a gunshot wound. *Courtesy of University of Melbourne Archives and the Vera Scantlebury Brown collection, 1984.0082*

She was closely mentored in surgical technique, and her letters vividly demonstrate her progress. She graduated from a nervous assistant and "clumsy – fingers all thumbs" surgeon dealing with her first case of gunshot wounds in May 1917, to blithely listing her day's activities 18 months later: tying off a femoral artery; repairing a damaged shoulder joint; repairing a gunshot wound to a hip joint; performing a secondary amputation; and setting a fractured scapula. The maturity and experience of the young surgeon from Australia were not to be underestimated.

For surgeons at the time – both male and female – such experience of complex injuries was by no means common. The weaponry used at the front often caused long bone fractures with extensive soft-tissue damage – a high proportion of casualties arriving at Endell Street were orthopedic cases. Later in the war, such patients were transferred to specialist orthopedic hospitals, but in the early stages the staff at the hospital saw a large number of patients requiring amputations and became proficient at making and fitting prosthetic limbs. Major abdominal surgery was often required and head injuries were common; particularly in the early years of the war.

Some patients required a craniotomy – or operation on their skull – to extract bullets or elevate depressed skull fractures. On admission, many patients went straight to the operating theatre, with around 20 operations performed daily.[26]

Vera kept notes on some of her cases. She corresponded regularly with Frank Kingsley Norris, her unofficial fiancé back in Australia, and compared cases with her brother, Cliff, who was posted to the Australian 29 Casualty Clearing Hospital in France. These notes record both her own anxieties as a surgeon, and illustrate the types of surgical cases:

> *I wish I knew more about knees. I have a man with a small bullet in the joint – I think so, X Ray not back. In 3 weeks swelling less – though some fluid no temp. I think rest and masterly inactivity the treatment. Cliff says they do not <u>drain</u> knees in France even when pus is in them. At Endell Street I saw opened wound washed out with saline then spirits then B.I.P paste then closed. Cliff says they use formalin and glycerine 20% in France and close. They are not like ours – pneumococcal joints at the C.H. which need drainage.*[27]

By May 1917 she was in charge of two surgical wards and performing more complex operations as primary surgeon, rather than assistant.

> *2nd patient GSW – removed the sequestra – cleaned up with BIP and put in two drains of gause very lightly. Very sick. Pulse poor vol.*[28]

---

[26] Murray, Flora, *Women as army surgeons*, London, Hodder and Stoughton, 1920, p. 69.
[27] Scantlebury-Brown Papers loose-leaf notes undated p.8 A6.
[28] ibid.

In addition to dealing with wounded soldiers, Vera and the other doctors also had to deal with deserters, malingerers, prisoners and drunkards (allocated to the aptly named 'Johnny Walker' Ward). Treating deserters required compassion and maturity: they were treated with disdain and rejected by society.

Given the horrors of the Western Front, some men were so desperate not to return to the war that they resorted to self injury. Maligned by their colleagues and wider society, some men went to extreme measures – speaking volumes about the severity of psychological trauma they faced. Some injected petrol into their knees or other joints, or soaked their limbs in candle grease and set them on fire. Desperate men exposed themselves to mustard gas – impossible to prove as self-inflicted, but deadly in its ability to choke the lungs, burn and blister the skin and to cause permanent blindness.[29]

Vera's letters also document life at Endell Street as part of a community, with themes of radical feminism and female fraternity. Her mentor – Dr Anderson – was committed to social reform and gender equality: before the war – in March 1912 – she had been arrested and sentenced to six weeks' hard labour for smashing a window in Knightsbridge during a suffrage protest.

For Vera, however, the dominance of suffrage as a topic of debate was tiresome; she was not alone in her views. Eleanor Bourne, a fellow surgeon from Queensland, and Emma Buckley, from Sydney, shared the view of many Australians that suffrage was no longer an issue. Australian women had been given the vote in 1902. In England however, former suffragettes remained committed to their cause, believing that their wartime actions would earn them their vote as soon as the war was over.

The hospital's feminist culture had its advantages: Vera enjoyed the independence that her new role provided, both from a personal and professional perspective. This newfound independence strained her long-distance relationship with Frank Kingsley Norris (her unofficial

---

[29] Stanley, Peter, 'Working the Nut: malingering and shirking', *Bad Characters: Sex, crime, mutiny, murder and the Australian Imperial Force*, Pier 9, Murdoch Books Pty Ltd, 2010, pp. 66–69.

fiancé) and her difficulty in deciding between the two is evident as she asked herself:

> Shall I give up medicine when I marry? Yes-No-Yes-No-Yes-No-all the days of life.[30]

In the end, her career won out. She renewed her contract at the hospital every six months and remained in England; "the silken cords of honour" keeping her there for however long the war continued. Soon, however, her relationship with Kingsley was over.[31]

The Endell Street Military Hospital remained open until December 1919. During that time its doctors saw 26,000 patients and performed over 7,000 major operations. Despite its contribution to the war effort, Endell Street remains largely forgotten. Often dismissed as a "voluntary hospital", its significance has been overlooked: its employment of women was progressive; its staff were committed to their work, despite the horrors they faced; and the "volunteers" were subject to the same regulations and conditions as any other hospital established by the War Office. Today the buildings remain and are home to an exclusive private members club. Its name, 'The Hospital Club', is the only indication of the building's wartime history.

THE WESTERN FRONT

Anderson and Murray were not the only women doctors to be frustrated in their attempts to offer medical services to the Imperial Forces. As the war continued, the Allies' need for medical care became desperate: women who could organise themselves, fund themselves and provide their own equipment saw an opportunity to contribute.

During the early years of the war, a number of all-female medical units working in auxiliary hospitals emerged, offering treatment near the Western Front. Supported by networks that originated in the suffragette movements, they were funded by donations and fundraising campaigns in

---

[30] *The Scantlebury-Brown Papers*, 29 June 1917, Vol. AN1, p. 12.
[31] Sheard, p. 98.

Britain and its dominions overseas, including Australia. Many well-known and well-connected women supported these units, seeing in them as an opportunity to advance women's emancipation by demonstrating their capability, skills and willingness to serve.

As was the case at Endell Street, these units encountered horrific injuries amongst the casualties they treated and were subject to strict regulation: the auxiliary hospitals treating both French and British officers and men were required to pass 'inspection' by RAMC authorities.[32] Working conditions were tough. Hospitals had to move frequently as ground was gained and lost, and to evacuate the wounded. Surgery was performed under almost impossible conditions, often during bombardment of surrounding areas. The majority of wounds were compound fractures and "almost all horribly septic", requiring the removal of dead and infected tissue (known as debridement) and fixation. Sanitation was primitive and water had to be boiled for surgical use on oil stoves.[33]

The auxiliary hospitals were required to be independent and could not rely on Allied military support for re-supply, accommodation or catering. They had to provide their own ambulances and drivers, as well as cooks, cleaners and orderly staff. These posts were willingly filled by women volunteers. Usually from privileged backgrounds and determined to support the war effort, many of the women had never worked before. They rose to the challenge, rolled up their sleeves and learned new skills under the toughest conditions.

One unit in Antwerp, in Belgium, was a British Field Hospital. It had a relatively large nursing staff and a team of doctors, with four men and four women. One of these women was Laura Foster, an Australian. Foster had obtained her medical degree in Berne in 1894 and had served in the First Balkan War as a nurse (women were not accepted as doctors during that campaign).[34]

Three days after the unit arrived it received an influx of casualties – 140 wounded soldiers in just 24 hours. As one witness recorded at the time,

---

[32] Murray, p. 4–6.

[33] Leneman, p. 164.

[34] Leneman, p. 164.

"although we had some very severe injuries, only about six or seven of the 140 died".[35] When Belgium fell to German forces, staff were forced to evacuate, together with their casualties. "It was a rather trying business loading the buses, the shells were landing thick and fast; one hitting a part of the hospital before we left."[36]

Dr (Majeure) Helen Sexton

Another intrepid Australian doctor followed a similar course and funded her own hospital in France. Dr Helen Sexton was the third woman to graduate from the Medical School of the University of Melbourne in 1892. She helped found the Queen Victoria Hospital for Women and Children and was one of the first female honorary surgeons in Australia.

Helen was more than 50 years old when war broke out, having largely retired from medical practice and moved to Europe in 1911 for "health reasons". She offered her services to the British Army and, like other female colleagues, was rebuffed. Undeterred, she decided to set up her own unit, which became a tented hospital at Auteuil near Paris. Financial support came from many of her Melbourne colleagues, and her efforts were widely reported in the Australian press. The hospital was visited by Australian officials and accredited by the French Government as a military hospital. She was given the rank of Majeure (Major). 'Majeure Hélène' clearly had good command of the French language and was well respected. In December 1915, it was reported in the Australian press that she had been given the honour of presenting medals to a French soldier. (The *Broken Hill Miner* incorrectly reported that Sexton herself had been awarded the Medaille Militaire and Croix de Guerre.)

Later in the war, Helen went on to hold a temporary appointment at the military hospital of Val-de-Grâce in Paris.[37] The Val-de-Grâce was a pioneering centre of facial plastic and reconstructive surgery and Helen

---

[35] Bellefaire, and Graf, p. 164.

[36] ibid, p. 165.

[37] Russell, Penny, 'Sexton, Hannah Mary Helen (1862–1950)', *Australian Dictionary of Biography*, http://adb.anu.edu.au/biography/sexton-hannah-mary-helen-8389/text14729, Accessed 19 September 2012.

encountered a large number of soldiers disfigured by their injuries. The hospital specialised in treating extreme cases, with many soldiers spending years undergoing multiple operations in an attempt to reconstruct shattered jaws, eye sockets or noses destroyed by shrapnel. Many were unable to talk and struggled to deal with their appearance; their trauma was exacerbated by the reactions of those who saw them.

It was a time of surgical experimentation, and surgeons subjected their patients to multiple operations in an attempt to re-create their features and their identities. Although huge advances in plastic and reconstructive surgery were made at the time, contemporary drawings and photographs cannot hide the pain and misery evident in what remained of their faces, despite the best efforts of Dr Sexton and her colleagues.

THE SCOTTISH WOMEN'S HOSPITALS

The largest, and best known, of all voluntary women's organisations during the Great War were the Scottish Women's Hospitals, run exclusively by women and established by Elsie Inglis, a Scottish doctor and campaigner for women's rights.

Inglis was inspired by her experience in setting up the first Medical College for Women in Edinburgh. She was also involved in the National Union of Women's Suffrage Societies (NUWSS); her idea was developed in direct response to attempts by the medical profession to exclude medical women from war service. Like many women doctors she had offered her services at the beginning of the war to the British War Office, but was firmly told she was not wanted. When she proposed to offer them a fully functional hospital, staffed by women, she was rebuffed with the advice: "My good lady, go home and sit still!"[38]

In Britain, female-run hospitals had a precedent, although not in a military context. Workhouse hospitals for women and children were largely run by women and care for the poor was a need met primarily by Women's Suffrage Societies. The appalling conditions encountered by

---

[38] Dr Elsie Inglis, 1864–1917. Remark quoted in Leneman, L., *Elsie Inglis. Founder of battlefield hospitals run entirely by women*. Edinburgh, National Museums of Scotland, 1998: p. 34.

Britain's 'public health' pioneers at the turn of the century went some way to preparing them for hospital management during the war.

In Australia, in 1897, the Queen Victoria Hospital opened in Sydney courtesy of funding from the 'Queen's Shilling Appeal'. Its mandate was, specifically, to be a 'Woman's Hospital run by Women'. The only distinction between this establishment and the women's hospitals of World War I being that the patients would also be women, not injured (male) soldiers.

Elsie Inglis was a woman of remarkable fortitude and persistence. Sit still she did not. Under her leadership the Scottish Women's Hospitals (SWH), supported by the NUWSS and funds raised by supporters in the United States, India, Australia and New Zealand, provided 14 medical units on foreign soil – many of them simultaneously and some in place for three to four years.[39]

The name – Scottish Women's Hospitals – was chosen to commemorate the Scottish origins of the scheme and to play down its suffragette origins, given antipathy towards the women's rights movement in some sectors of society. The SWH remained independent throughout the war and operated on a number of different fronts. Although 'Scottish' in name, they were administered from London and their staff included women from England, Wales, Canada and Australia.

The hospitals' infrastructure and expertise was offered to the British Army, but given conservative attitudes at the time, the offer was refused. The SWH received a different response from the armies of France and Serbia: both were in desperate need of medical support. With an inadequate supply of surgeons and hospitals, the French and Serbian armies welcomed the assistance of female doctors, particularly as they arrived fully staffed and with sufficient equipment and stores.

Notable Australian women who served with the SWH were Dr Agnes Bennett, who commanded a SWH unit in Macedonia from the summer of 1916 until autumn 1918, when the unit moved to Vrnaja in Serbia;[40]

---

[39] Mitchell, A. M. p. 11. Authors' note: Dr Inglis originally sought £22,000, estimating a need for 100 beds, four doctors, 10 nurses and two cooks (with salaries initially waived). In fact, appeals raised £449,000, largely due to overseas donations.

[40] Leneman, p. 167.

and Dr Laura Hope (née Fowler), who was the first woman to graduate in medicine from Adelaide University in 1891. Accompanied by her husband, Charles, who was also a doctor, the couple was captured and imprisoned in November 1915 when Serbia was overrun by Austria-Hungary.[41]

Other notable women were Dr Lilian Cooper who, accompanied by her companion, Miss Josephine Bedford, served with the SWH in Greece before becoming a frontline surgeon during the Macedonian campaign; and Olive King, who was a young ambulance driver, also during the Macedonian campaign. She ferried casualties to and from hospital before resigning to join the Serbian Army. Each of these women received foreign decorations in recognition of their bravery and service.

ROYAUMONT

Back on the Western Front, conditions on the battlefield deteriorated. In 1916, the military situation had become particularly grim with repeated attacks by both sides, enormous losses, but barely any change in the ground won or lost. In the trenches conditions were particularly appalling. Casualty losses were high from bombardments, from senseless forays into the sights of enemy guns, and from disease. On one day alone, on 1 July 1916, the British lost 60,000 men: 20,000 were killed and 40,000 were wounded. German bombers paralysed the French with their latest weapon – deadly poison (diphosgene) gas. French losses were enormous and their medical situation desperate. So began five months of terrible and pointless slaughter – the 'Great Push'.

In this hellish environment, the SWH continued their work. A key medical unit was the SWH established in the medieval abbey of Royaumont, around 30km north of Paris. Set in rolling wooded countryside, the hospital remained open from January 1915 until March 1919. In that time it grew from 100 to 600 beds and became the largest voluntary hospital in France to operate continuously throughout the war.

The transformation of the ancient abbey into a fully functioning

---

[41] Sandford Morgan, Elma, *A Short History of Medical Women in Australia*, Melbourne, 1970 8–9; *BMJ*, 1916, I, p. 288.

Operating theatre at Royaumont (unknown surgeon). *Image source: Crofton, Eileen,* The Women of Royaumont, *Tuckwell Press, Scotland, 1997*

hospital was no small achievement. Unused for nearly 100 years, the obstacles were immense. In his book, *The Scottish Women's Hospital at the French abbey of Royaumont*, published in 1917, author Antonio de Navarro describes the enormity of the task. At first, there was no lighting other than candles and water was heated on a single stove in the middle of the room.

> All day long, the nurses, orderlies, junior doctors – healthy, active young women, with a full share of the modern girl's strength – scrubbed, dusted, washed the floors and walls, opened huge packing cases and carried up flights of stairs their heavy contents: beds, bedding, and all the paraphernalia necessary for fully equipped wards.[42]

Close to the front line, Royaumont was set up as a casualty clearing station capable of performing wound surgery on casualties taken straight from the front. In 1917, a small 'satellite' hospital was also established

---

[42] de Navarro, Antonio, *The Scottish Women's Hospital at the French abbey of Royaumont,* 1917. Reprinted from the Collection of the University of California Libraries, p. 103.

at Villers-Cotterêts. At both locations, women doctors developed surgical skills that proved to be invaluable during the Battle of the Somme in 1916 and the rapid German advances of 1918. In this era, before the introduction of antibiotics, meticulous surgical debridement of injuries with repeated daily dressing of wounds was essential. At Villers-Cotterêts, operations were carried out by candlelight until they were forced to abandon the site to advancing Germans. The staff who served here (including an Australian, Millicent Armstrong, who is mentioned later in this chapter) were awarded the Croix de Guerre.[43]

In all, 10,861 patients were admitted to Royaumont and its ancillary hospital at Villers-Cotterêts. Of those patients, 8,752 were soldiers from the French and French Colonial Armies; others were British or American. There were also German prisoners-of-war. The death rate amongst servicemen was 1.82 per cent, a remarkably low figure given the severity of the soldiers' injuries.[44] Royaumont was one of the few centres chosen by the Institut Pasteur for studies in the treatment of gas gangrene.

Today, surrounded by forested park and clusters of wild blue cornflowers, the serenity of the abbey belies its turbulent past. The Royaumont Monument hidden in the grounds marks the limit of the German offensive in 1914; it also commemorates the lives of those who died at the hospital.

At the entrance to the abbey there is a single plaque. There is also a tree planted by Her Royal Highness, The Princess Royal, as a tribute to the 'Scottish' women. This is all that remains to record the endurance, hardship and danger encountered by these women during the war. There is little other sign of their professional service or its legacy.

Dr (later Captain) Elsie Dalyell

One of the distinguished Royaumont 'doctoresses' was an Australian – Dr Elsie Dalyell. Born in Sydney in 1881, Elsie initially trained as a teacher.

---

[43] Leneman, p. 167.

[44] Bowden's foreword to Crofton, Eileen, *The Women of Rouyamont*, Tuckwell Press, Scotland, 1997.

After the 'shattering blow' of a hysterectomy and a failed romance she decided, however, to study medicine.

As a student she was said to be an 'extremely attractive girl with corn-yellow hair, a fair complexion and blue eyes'.[45] Her looks, intellect and reputation as a fast motorbike rider (risqué for a woman, at the time) led her colleagues to call her 'the Yellow Peril'. She graduated from Sydney University in 1909 with first class honours, and became the first woman to be appointed to the full-time staff of the medical school as a demonstrator in pathology.

When the war broke out she was working in Britain – she was the first Australian woman awarded a coveted Beit Research Fellowship at the Lister School of Preventive Medicine in London. Like other female doctors, her offer of service to the War Office was rejected. Instead, she joined Lady Windborne's Serbian Relief fund, hoping to be sent with them to Skopje to help with the typhus epidemic that had devastated Allied troops in 1915. Instead, she found herself safely located in Addington Park, Croydon. In search of adventure, she decided to join the Scottish Women's Hospitals and was immediately sent to Royaumont in France.

Elise was appointed as a bacteriologist. As one of only two doctors available, and facing a continuous flow of patients with all types of infectious diseases, she also turned her hand to more general medical tasks. She took charge of the bacteriological laboratory and served there from May to October 1916. The focus of her work was the infection of war wounds, and in particular, the dreaded effects of gas gangrene.

Gas gangrene was a deadly illness, predominant during the Great War. Few doctors were familiar with its symptoms or treatment. Even today, gas gangrene is still feared for its speed of progress, its resistance to antibiotics and the possibility of a return to the ferocity and fatality rates of the past. Unique in its characteristics, gas gangrene is the result of complicated bacterial processes. Thriving on dead and damaged muscle, multiple organisms work together in a deadly combination to form gas within the tissues; a crackling sound in the tissues (crepitation); discoloration

---

[45] Richardson, G. D., 'The Dalyells and their Kin' (1988) quoted in *The Women of Royaumont* p. 267.

of muscle; swelling; and foul-smelling fluid that exudes from wounds. The organisms involved produce powerful toxins, similar to those of the botulism toxin, which combine to destroy nearby tissue. The gas separates and opens new spaces in the tissue for the bacteria to enter.

Death follows rapidly, but unlike many other infectious diseases, the patient does not lose consciousness until close to death.

> The most terrible of all the horrors which come under the care of the surgeon in this war is undoubtedly gas gangrene. Dramatic in the suddenness of its onset, the rapidity of its advance, and the repulsiveness of it too frequently fatal outcome, it has reaped a cruel harvest of our young and vigorous manhood.[46]

At the time, Professor Michel Weinberg of the Institut Pasteur in Paris was the leading authority on the bacteriology of gas gangrene infections. The illness was particularly common in France given the large amounts of manure used on the land: the battlefields contained high concentrations of the bacteria responsible.

This, combined with the mutilating injuries caused by high-explosive shells and machine guns, and wounds contaminated by clothing, soil and metallic fragments, created an ideal environment for the disease to flourish and do its worst.

Dr Dalyell and her colleagues at Royaumont advanced understanding of this dreaded disease. Their contribution to the development of serum to counter gas gangrene is, perhaps, their greatest legacy. Used widely during the Great War, and sporadically in World War II, the treatment has now fallen out of use, partly due to advances in antibiotics and debridement techniques (the removal of damaged or infected tissue).[47]

As the casualties arrived from the battlefield, bacterial smears were taken from every wound and sent to the laboratory for examination. Elsie would stain them meticulously, initially dipping the slide in crystal violet, then heat fixing each slide over a Bunsen burner before adding a counterstain.

---

[46] Savill A., *Archives of Radiology and Electrotherapy*, Vol. XXI, No. 7, December 1916.

[47] Schraibman, I. G.,' Antiserum in gas gangrene', *British Medical Journal*, 1968, March 16, vol. 1(5593), p. 704.

Looking under the microscope she could identify the tiny purple rods characteristic of the illness, sometimes with their spores resembling miniature purple tennis racquets. The presence of these microscopic enemies was reported back to the ward, even before the final identification of the organisms took place. Smear films and culture fluid, stored in small individually labeled bottles of broth, were then sent to the Institut Pasteur.

The worst cases were treated with anti-serum, usually in the operating theatre while the patient was anaesthetised and the wounds were debrided. Ruthless debridement was the rule involving amputation, or excision of all dead or partially dead muscle and the thorough cleaning and removal of all foreign material, shell fragments, clothing and fragments of bone. Even the smallest amount of dead tissue left in a patient would be an invitation for the dreaded organisms to continue to flourish.

In these early days anti-serum was scarce and reserved only for the most severely infected cases. The first publication from Royaumont, detailing treatment of the illness, clearly describes the different types of gas gangrene encountered. It also describes the successful treatment of 5 out of 10 patients; before treatment, all were expected to die.[48]

The first week of the 'Great Push' in July 1916 proved to be Elsie's most difficult week. Royaumont was situated just 40km from the Front and casualties reached the hospital within hours of being wounded. Others, if they were delayed at casualty clearing posts, took several days to arrive following injury:

> The incessant thunder and boom of the great guns had never been silent for days. This day, at dawn, the thunder had swelled to an orgy of terrific sound that made the whole earth shiver; then, a few hours later, had ceased, and we could hear once more the isolated reports of individual cannon … Trains were arriving from the Somme in one long stream. The drivers never ceased journeying backwards and forwards all afternoon and all that night, and the three women and the man, who drove our four

---

[48] Ivens, F. M., 'A Clinical Study of Anaerobic Wound Infection. Analysis of 107 cases of Gas Gangrene'. *Proceedings of the Royal Society of Medicine 1916–1917*, X Part 3, Surgical Section, pp. 29–110.

ambulances, carried over a hundred cases over the first 24 hours of that nightless week ... during the first twenty days of the Great Push there were always some of our cars at the clearing station day and night ...

Their wounds were terrible ... many of these men were wounded – dangerously – in two, three, four and five places. That great enemy of the surgeon who would conserve life and limb, gas gangrene, was already at work in 90% of cases. Hence the need for immediate operation, often immediate amputation. The surgeon did not stop to search for shrapnel and pieces of metal: their one aim was to open up and clean up the wounds, or to cut off the mortifying limb before the dreaded gangrene had tracked its way into the vital parts of the body. The stench was very bad. Most of the poor fellows were too far gone to say much ...[49]

During this time Elsie worked daily in the laboratory but, as one of the doctors, she was also expected to help in the operating theatre, where she gave ether anaesthetics. Her recollections of those first few days during the 'Push' are recorded in a letter to the *Sydney Telegraph*:

There have been simply awful days and nights. I hear the poor people in the operating theatre cleaning up madly ... a swab of every wound is taken immediately after admission and sent instantly to the laboratory. I examine them on the instant – did 180 in three days – and a bacteriological report was despatched within about half an hour. All the gas gangrene cases were sorted out by this means and the incipient cases were spotted at once and operated according to the severity of the infection as notified by me. By that means we saved lots of limbs, as they were spotted early and opened up in time. So many of the cases I insisted on immediate operation proved to have early gas formation at the bottom of a deep narrow shell wound that it would certainly be fatal to leave them; they would probably have lost a limb, if not

---

[49] Crofton, Eileen, *The women of Royaumont: a Scottish Women's Hospital on the Western Front*, East Linton, Tuckwell Press, 1996 p. 69.

life. So far we have had only 5 deaths and we did 112 bad cases of gas gangrene without stopping. As I did the bacteriological work in batches of 20 …, I was able to take a hand in the theatre and plied between the lab and the operating table without ceasing day or night.[50]

Dr Dayell was a popular member of the Royaumont team. She was remembered for her 'frame shaking chuckle', her friendships, her courage, compassion, common sense and humor. One of her Royaumont colleagues, Dr Martland, wrote:

> … she certainly became a ubiquitous and beneficent presence throughout Royaumont. If a medical officer wanted to give an anaesthetic or evacuate a batch of blessés, there was Dalyell, a calm, fair, massive figure, always available, utterly efficient, never in a 'flap'. She had genius for appearing in any place where trouble was; that soft Australian voice murmuring, 'Can do, honeyeee?' (sic) in a tight spot, is one of my most blessed memories of Royaumont. Another good memory is of her diverting a bunch of depressed 'doctoresses' by quick-change impersonations of Scottish women in the incredible variety of tartan-trimmed garments which Edinburgh considered suitable as a uniform. Her gaiety was one of the best gifts to Royaumont. I never saw her gloomy, though sometimes saddened by the waste and muddle of war. Just to speak of Dr Dalyell brings a sense of healing to the spirit – that is the kind of woman she was.[51]

Following her service in France, Elsie volunteered for the RAMC, serving as a bacteriologist in Malta and Salonika. Although the RAMC was now accepting women doctors, they were only 'attached': they still carried no real rank or status.

Early in 1919, Elsie went to Constantinople to offer her expertise during a cholera epidemic. In June the same year her services were recognised with

---

[50] Dr Elsie Dalyell. Letter to the Sydney and NSW *Daily Telegraph* 22 January, 1918 (quoted in Crofton, p. 75).

[51] Crofton, pp. 269–70.

the award of an OBE. She was the first Australian woman doctor to be decorated for her services to military medicine within the British Honours system; she was recognised as a civilian, however, and not as a member of the military. Even her Record of Service (one of only a few records surviving) fails to record two occasions, in 1918 and 1919, when she was Mentioned-in-Dispatches for her contribution.[52]

Miss Millicent Armstrong CdeG

One of the Australians who served at the Villers-Cotterêts detachment was Millicent Armstrong. Known for her artistic temperament as well as her bravery, Millicent worked as a volunteer medical orderly at Royaumont. Some historians have referred to her a nurse, although she had no formal nursing training.

Like other volunteer orderlies, Millicent was not paid. Many came from the so-called 'gentle classes', but it is doubtful whether gentility prepared them for the tasks ahead. Their role was to undertake menial work, including cooking, scullery duties and taking turns as the 'night hall porter' – waking other staff when convoys of wounded arrived. They also undertook laundry duties; cleaning, disinfecting and repairing damaged uniforms, along with their owners.[53] Sometimes they would help out with X-rays, or even in the operating theatre. They also read to patients, helped them write letters and provided much needed recreation.

Millicent had graduated from Sydney University with a Bachelor of Arts and First Class Honours degree in English. She had an artistic temperament and organised events at Royaumont to boost the morale of staff and patients. Writing her own material, and creating her own props, she put on pantomimes, melodramas and variety shows, offering a brief escape from trauma and the horrors of war.[54]

---

[52] Mitchell, Ann, M., 'Dalyell, Elsie Jean (1881–1948)', *Australian Dictionary of Biography*, http://adb.anu.edu.au/biography/dalyell-elsie-jean-5575/text9995 Accessed 28 Jan 2012.

[53] See de Navarro, chapter VIII 'Orderlies' and Chapter X 'Vêtements'.

[54] Blackmore, Kate, 'Armstrong, Millicent Sylvia (1888–1973)', *Australian Dictionary of Biography*, National Centre of Biography, Australian National University, http://adb.anu.edu.au/biography/armstrong-millicent-sylvia-9385/text16489, accessed 24 November 2012.

Millicent's concerts invariably ended with a gramophone rendering of the Marseillaise and rousing reply of 'God Save the King'. [55] She subsequently wrote a number of plays under her nom-de-plume, Emily Brown, some of which were mildly successful. Regrettably, none of her wartime writing has survived.

Millicent was also part of a small group of women deployed to Royaumont's outpost at Villers-Cotterêts in May 1918. As a result of her actions in the face of the advancing German offensive she was awarded the Croix du Guerre, France's highest honour, for her bravery in rescuing wounded soldiers while under fire.

THE SERBIAN CAMPAIGN: A FORGOTTEN WAR

On the border between the now Republic of Macedonia and Greece sits Mt Kajmakcalan, surrounded by imposing peaks. It was here, during World War I, that extraordinary feats of military surgery were undertaken by women.

The doctors involved included Australian women and a New Zealander, Dr Jessie Scott. Australian staff also included Miss Josephine Bedford (mentioned above) and Olive King, who served as ambulance drivers; Miles Franklin, later the author of *My Brilliant Career*, volunteered as a medical orderly.[56] Sister Agnes Kerr worked as a nurse, and tragically died during the campaign from the effects of malaria.

These women all worked under the leadership of Dr Agnes Bennett, an Australian who was born in Sydney and educated in Britain, who established a successful medical practice in New Zealand before the war.

Captain Agnes Bennett

Aged 42 when war broke out, Dr Bennett was a woman with considerable professional experience and an established medical practice in New Zealand.

---

[55] See: de Navarro, chapter XVII, 'Recreation'.
[56] The stories of Olive King, Josephine Bedford, Agnes Bennett and Lilian Violet Cooper are beautifully recounted in Susanna De Vreis's *Heroic Australian Women in War*, Harper Collins, 2004.

The war in Serbia featured prominently on recruitment posters during World War I.

Her early career revealed a woman with single-minded determination. She took out a bank loan to pursue a university education and paid her own way to study medicine at the Edinburgh University Medical College, one of the first medical colleges in Britain to admit women. There she was mentored by Elsie Inglis, a woman who became a key influence in her decision to go to war.

Agnes found it difficult to carve out a career as a doctor in Britain. She wanted to pursue a career in surgery, but unsurprisingly this proved to be impossible. Residency jobs in surgery were much prized and women were not considered, even if their examination grades were superior. Surgery remained a bastion of male privilege.

After a frustrating search for a suitable (or indeed any) post, Bennett was given a position at the Larbert lunatic asylum – a Dickensian institution where she worked for 15 months.[57] She fared no better on her return to Australia. Unable to find a suitable post and needing to pay back her bank

---

[57] De Vries. *Heroic Australian Women in War*, p. 76.

loan, she accepted a position at Callan Park Hospital for the Insane in Sydney.

Soon, however, she was offered a position in Wellington, which she accepted. She established a career at Wellington's St Helen's Hospital as the Chief Maternity Officer and became one of New Zealand's pre-eminent woman doctors. Throughout these years her dreams of practising surgery were still very much alive and she went back to Edinburgh for further postgraduate study. She completed her degree in surgery in 1911, just a few years before the war broke out.[58]

When her three brothers enlisted for service with the Australian Imperial Force (AIF) Agnes returned to Sydney with the intention of joining them. When she offered her services to the Australian Army, however, she was rejected. The recruiting medical officer told her that if she wanted to help, she would be "better to go home and knit". Accustomed to rejection on the grounds of her gender, Dr Bennett was furious all the same. She wrote to her friend and mentor Dr Inglis, who by this stage was setting up the first of the Scottish Women's Hospitals.

Determined to serve in a medical capacity, Agnes paid her own passage to England, departing in mid-1915. Originally, she intended to work with Helen Sexton (mentioned above). Dr Sexton was a doctor from Melbourne who had set up a tented hospital in Auteuil in France funded by private donations.

Agnes's plans changed, however, when her ship docked at Alexandria. Confronted by the sight of rows of wounded men waiting on the docks in the heat of the sun, she realised that these were ANZAC soldiers – men just like her brothers. She made a snap decision. Like most surgeons of the time, Bennett was travelling with her own surgical instruments. She immediately offered herself and her instruments to the commander of medical services in Alexandria.[59]

Agnes was attached to the New Zealand Medical Corps in Egypt at Pont

---

[58] Ann Curthoys, 'Bennett, Agnes Elizabeth Lloyd (1872–1960)', *Australian Dictionary of Biography*, National Centre of Biography, Australian National University, http://adb.anu.edu.au/biography/bennett-agnes-elizabeth-lloyd-5206/text8761, Accessed 29 May 2013.

[59] Gilchrist, Hugh, 'Australians in Macedonia: the Women', *Australians and Greeks Vol II*, Halstead Press, Sydney, 1997, p. 132.

de Koubbeh and Choubra, becoming reportedly the first woman surgeon to be appointed in any armed service of Britain and the Empire.[60] Her rank was Captain. The hospitals in Cairo and Alexandria were understaffed with overcrowded wards. The wounded came from Gallipoli; the failed assault on the Dardanelles. Although they had survived the horrors of the campaign, their lives were still in the balance, threatened by gangrene or severe injuries.

Typhoid, a scourge of many wars, was rampant. Many young men who were evacuated to Egypt died, not from battle wounds, but from disease. Spread in contaminated water, the early symptoms of typhoid can be vague, with abdominal pain and diarrhoea. Within days the fever and diarrhoea worsen. Typically, typhoid victims have ragingly high fevers (over 39.5 degrees Celsius), which may be accompanied by delirium and hallucinations. Perforation of the bowel and peritonitis are common and almost universally fatal. For others the disease is marked by the appearance of 'rose spots' on the abdomen and chest – in the pre-antibiotic era, these were usually a harbinger of death.

Captain Bennett demonstrated her characteristic determination and remained resourceful, despite the appalling conditions. When there were not enough beds she requisitioned a car, ripped up bunks from laid-up tourist steamers on the Nile and used them for the cause.[61] She also seized the opportunity to develop her medical skills as she had dreamt of for many years.

When the Anzacs were diverted to the battles on the Western Front, Captain Bennett followed them to Europe, arriving in London in early 1916. She met with Dr Inglis, her former mentor, and offered her services to the Scottish Women's Hospitals. Dr Inglis, knowing of her character and talents both as a surgeon and an administrator, and her recent experience in Egypt, offered her command of a new field hospital. Set up just behind the front line in Greek Macedonia, the hospital would handle casualties from the Serbian Campaign, where Britain's Allies – the Serbs and the

---

[60] Manson C. and Manson C., *Dr Agnes Bennett*, Michael Joseph, London, 1960, p. 72; see also *Australians and Greeks*.

[61] De Vries, *Heroic Australian Women in War* p. 82.

French – were fighting German-backed Austrian troops.

Regarded as something of a 'side show', compared to the numbers of troops and casualties on the Western Front, this was a site of fierce fighting. Aided by German aircraft, Bulgarian bandits engaged in hand-to-hand combat in the mountains, hijacking lorries and showing no mercy to the Serbs or their French allies.

By 1916, the Serbian Army had taken heavy losses. Devastated by Bulgarian forces, Serbs had lost large amounts of territory and had little in the way of reinforcements. Soldiers from the Irish Fusiliers and Britain's Black Watch regiment were sent to help them and a large British field hospital was set up in Salonika. Still, the Serbs remained without adequate resources. They had no hospitals, few ambulances and no trained staff.

Dr Bennett showed no hesitation. Released from duty with the Commonwealth forces she signed on as a senior surgeon. She was duly appointed Commanding Officer of the new SWH facility with a salary of £200 per year plus expenses. By this stage of the war, Dr Inglis had hospitals located in France, Serbia and Russia. Control was a challenge and she relied heavily on the skills, experience and judgment of her Commanding Officers. In London, Dr Bennett used her time to arrange the necessary stores and staff. A number of other Australians would work with her as doctors and nurses as well as ambulance drivers and volunteer orderlies. Her surgeons included the New Zealander Dr Jessie Scott and Dr Lilian Cooper from Queensland, who arrived with her long-term companion, Miss Josephine Bedford.

Dr Bennett was also allocated several Australian nurses, including Caroline Reid (from Melbourne), Florence Grylls (from Western Australia), Robbina Ross and Mary Stirling (from Adelaide). Miles Franklin joined later as a medical orderly and Olive King, from Sydney, also joined the unit as an ambulance driver, bringing her own privately funded ambulance, nicknamed 'Ella the Elephant'.

Despite its Scottish heritage, and although it was commanded by an Australian with staff from British Dominions, the new unit was known – somewhat confusingly – as America Unit No. 1, because the funds for its equipment and operation were raised in the United States. Its Scottish heritage was reflected in its uniforms – the women wore grey woollen coats with tartan lapels, heavy grey skirts and grey flannel shirts with a Gordon tartan scarf. The doctors were issued with peaked caps (similar to those of the British Army) while the nurses wore felt hats.

Staff were issued with ammunition and revolvers. The Bulgarians had a fearful reputation for rape and execution and this was a necessary precaution for the women and their patients. It was a move that proved to be well-justified. Hospitals were also in danger – a French field hospital had been attacked only months earlier and the throats of the staff were cut, sparing no-one.[62]

America Unit No. 1 set up its hospital on the slopes of a mountain, eight kilometres from the town of Ostrovo in northern Serbia, close enough to the railway station to facilitate the moving of casualties out of harm's way.

Despite the war, Agnes was not oblivious to the beauty of the place. In winter the snow-covered peaks, deep gorges and canyons were spectacular and worthy of any postcard. In summer, the same mountain slopes were covered in wildflowers and poppies, their colours reflected in the waters of nearby Lake Ostrovo.[63] She rapidly established the unit as a field hospital with 200 beds under canvas. For the operating theatre, she availed herself of a barn, with a small area portioned off for X-rays and a secondary operating theatre was established in a neighboring tent.

As at Royaumont, X-rays proved vital in locating pieces of metallic or bone fragment. Without finding and removing these fragments they could become a source of sepsis and gangrene. Dr Cooper, who would later join the Ostrovo team, had seen the first demonstration of X-rays in Queensland in August 1896, less than one year after their discovery by William Roentgen.[64] The early X-ray machines were considered essential pieces of equipment, but were cumbersome and radiation safety was not yet understood – this shortened the lives of many early users.

Within its first week of operations, 160 injured soldiers were brought to the Ostrovo hospital. In the first few days alone, more than 10 amputations

---

[62] De Vries, *Heroic Australian Women in War* p. 85.

[63] Lake Ostrovo is now known as Lake Vergoritis, located in northern Greece.

[64] Williams, Lesley, M., *No Easy Path. The Life and Times of Lilian Violet Cooper. MD, FRACS (1861–1947), Australia's first woman surgeon*, Amphion Press, Brisbane, 1991, p. 24.

were performed – an indication of the high workload that would soon follow. Battlefield amputation, whilst not technically complex, required strength of character, a measure of resolve and the self-assurance that such an act of mutilation is in the best interests of the patient. This was an era when surgical specialisation had not yet developed. Surgeons were expected to remove an appendix or amputate a limb with equal skill.

Casualties were brought down from the frontline over difficult mountainous terrain. The T-model Ford lorry ambulances were driven by women under treacherous conditions; they negotiated hairpin bends up the steep slopes of the Kajmakcalan range over dirt-covered roads covered with potholes. The drivers worked around the clock. Most made four to five trips per day, starting at first light; each journey involved a hair-raising round trip of 28 miles over rough roads made slippery from sleet and melting snow. Frequently the ambulances were bogged or broke down. One went over the side killing the driver. Many of the ambulance drivers were young women of private means who had never worked before, far less been allowed to drive a motor vehicle. Their pluck and adaptability proved to be worthy of any *Girl's Own Annual*.

The Serbian winter campaign of 1916–1917 was particularly cruel. The war conspired with the weather to create the harshest of conditions. The Third Serbian Army and its French allies were continuing to take heavy losses. The wards of Dr Bennett's hospital were overflowing with casualties and the winter only made matters worse. With the rains came trench foot. Soldiers' feet became soggy, swollen, bleeding pulps, fertile for infection. Boots were worn through, fresh socks unavailable and the constant marshy conditions only exacerbated the problem. Prolonged exposure to damp, unsanitary, cold conditions eventually took its toll. Open sores became infected, toes began to decay with gangrene and amputation was often the only option.

Conditions were harsh that long, bitter winter. Cold permeated everywhere. Orderlies brought snow-drifts into the wards as they delivered their patients on stretchers. Cold draughts found their way under the tent flaps. Wood was scarce and had to be scavenged from the mountain slopes, within earshot of the guns.

The cold weather brought hunger, as food became scarce. The Serbs were dying of starvation and soon the hospital also suffered from a shortage of supplies. What little they had brought from England by way of tinned rations, began to run out. Rations for the staff and patients were stretched by the resourcefulness of the locals, who would often boil the local mountain weed, known as 'horta' – as a substitute for fresh vegetables.[65]

As their fingers stiffened with the cold, the surgeons started operating in woollen mittens rather than rubber gloves. As the winter did its worst, the numbers of casualties increased and the Serbs' military situation grew desperate.

At the height of the winter campaign, on the 18th September 1916, Dr Bennett gratefully received two new arrivals from Australia – Dr Lilian Violet Cooper and her companion, Miss Josephine Bedford.

Dr Lilian Violet Cooper

Lilian Cooper became known as a pioneer female doctor in Queensland. She was also the first woman to be admitted as a Fellow to the Royal Australasian College of Surgeons.

Born in 1861 in Kent, Lilian was the third of eight children. Her father was an army officer, Lieutentant Colonel Henry Fallowfield Cooper of the Royal Marines Light Infantry, and her elder brother became a Colonel in the Royal Artillery serving in Burma and in World War I.[66]

She studied medicine at the London School of Medicine for Women, housed in an elegant, tree-shaded house near Brunswick Square, a short walk from the London School of Tropical Medicine and the British Museum. At that time, England still did not permit women to obtain medical degrees, so in 1890 she successfully undertook the joint examination of the Royal College of Physicians and Surgeons in Edinburgh and the Faculty of the Physicians and Surgeons in Glasgow. Later in her career, she spent time in both England and America undertaking post-graduate studies in surgery.[67]

---

[65] Gilchrist H.

[66] Williams, *No Easy Path* p. 1.

[67] Cazalar, L., *Dr Lilian Cooper*, Unpublished monograph in Archives of the Queensland Branch of the Australian Medical Association, 1970, p. 1. Cited in Williams, p. 29.

Dr Lilian Violet Cooper and Miss Josephine Bedford cleaning boots outside a tent, Serbia. *Image: Bennett, Agnes Elizabeth Lloyd, 1872–1960. Courtesy of the Alexander Turnbull Library, Wellington, New Zealand* (http://natlib.govt.nz/records/22333288)

Dr Cooper was the first woman to practice medicine in Queensland. When an opportunity presented itself in 1891, she emigrated with her lifelong companion Miss Josephine Bedford, to Brisbane aboard the *RMS Lusitania*.

No stranger to hard work, she often performed house calls on bicycle – late at night, wearing long skirts over harsh roads poorly lit by gaslight. She was also said to have performed an appendectomy on a young woman using the dining-room table for the operation and boiling her instruments in a saucepan over a wood stove. The woman's husband administered the anesthetic under Dr Cooper's close supervision.[68]

By 1911 she had a thriving practice and her reputation as a doctor was well established. She was keen to advance her own surgical skill, however, and learn new techniques. To do so she undertook a study tour of the United States and Britain. She attended the esteemed Mayo Clinic in Minnesota where she had the opportunity to observe one of the foremost

---

[68] ibid.

stomach surgeons of the time, Dr William Mayo. She also saw his brother at work, Dr Charles Mayo, who was recognised for his skill in head and neck surgery.

In England she had the opportunity to observe the work of Sir Berkely Moynihan, a noted British abdominal surgeon, and Sir Arbuthnot Lane, who was undertaking pioneering work using steel plates to fix fragments of fractured bone. She studied hard to improve her knowledge and obtained a higher degree (MD), which was awarded by Durham University.

Lilian had immense energy and a seemingly tireless capacity for work. She was a pragmatic woman who had worked hard to overcome the inevitable prejudice of her position. Described as a "tall, angular, brusque, energetic woman, prone to bad language", stories about her outspokenness, swearing and abrupt manner were legendary.[69] She had also, however, established a reputation as being meticulous, hard-working and possessing a deep sensitivity to the needs of her patients.

By 1914 she had made her way as a pioneer in medicine, despite opposition and prejudice, demonstrating remarkable strength of character:

> As a surgeon, the doctor displays a skill, a coolness and a celerity which is not readily understood by those who have not learned that some women can in an emergency, summon up a nervous force and will-power above that of the other sex, ...
>
> She has worn down all that barbaric opposition which once existed against her in her capacity as a lady doctor, by sheer good nature and hard work.[70]

By the outbreak of the Great War, aged 55, Lilian had established a successful and busy practice working in private and public hospitals. Still, she chose to respond to an advertisement by the Scottish Women's Hospitals in the *British Medical Journal*. She was to join Agnes Bennett

---

[69] Leggett, C. A. C., 'Cooper, Lilian Violet', *Australian Dictionary of Biography*, Vol. 8, Melbourne, 1981, p. 105. Available at http://adb.anu.edu.au/biography/cooper-lilian-violet-5770, Accessed 13 June 2013.

[70] *Queensland 1900 – A Narrative of Her Past, together with biographies of Her Leading Men*, compiled by the Alcazar Press, Brisbane, W.H. Wendt and Co Printers, Lithographers, Stationers and Engravers, Edward St., Brisbane, p. 175.

and spend the next year working as a frontline military surgeon during the Macedonian campaign, operating on the wounded of the Third Serbian Army.

Like Dr Bennett, Dr Cooper was offered an honorarium of £200 per year, plus expenses. Her companion, Miss Bedford, signed on as an orderly, receiving only expenses, but no salary.[71]

There is little doubt that Dr Bennett welcomed the arrival of the capable Dr Cooper and Miss Bedford. She later wrote to a friend:

> Dr Cooper is I think, happy. She looks fat and jolly to what she did on arrival and I think Miss Bedford is happy too organising all our motor transport. It will be 15 ambulances, two trucks and a touring car very shortly. It is a treat to have those two.[72]

Miss Bedford took control of the drivers and proved to be a capable organiser. But as the impact of the winter campaign intensified, the journey from the frontline became more precarious and patients were dying en route. Despite the risks, Dr Bennett decided the only course of action was to set up an advanced dressing station closer to the front. This would allow the most seriously injured patients to be stabilised and undergo surgery immediately.

Staffed by Dr Cooper and Miss Bedford, the Dobraveni dressing station opened a week before Christmas 1916. It was little more than a makeshift camp. The dining room roof comprised a tarpaulin. Water had to be carted from a nearby village, two kilometers away.

> The conditions were appalling by anyone's standards: They made their own mess-house from petrol cans; the kitchen was made by piling stones up in the fashion of the dry-stone walls of Scotland, the roof being hammered out cartridge cases. The wards and staff quarters were tents; temperatures were sub-zero. Wood was

---

[71] Personal Files of Dr Cooper and Miss Bedford: Agreement with Administrators and Medical Staff – Dr Cooper; Agreement with Unsalaried Employees – Miss Bedford, The Mitchell Library, North Street, Glasgow, Archives of the SWH, Tin 36.

[72] Agnes Lloyd Bennett MS1346 papers/176, Letter of 11/11/1916 to 'The Duchess', Alexander Turnbull Library, Wellington, NZ.

brought by mules from a mountain some kilometres away – a 5-hour trip in each direction. One sister recorded that often her hands were too cold to write and the tea froze in the cup if they were slow at drinking.[73]

The wounded were transported to the dressing station by mules. Usually two casualties were carried by each mule, seated in makeshift seats on either side. It was a rough journey, with no pain relief, each step jolting a fracture or jarring a chest injury. The more seriously injured were brought down by stretcher.

Dobraveni was a dangerous place. Located less than 10 kilometres from the frontline positions of the enemy, and within range of the guns, the only means of retreat was by foot – a walk of 20 kilometers down the mountain. Lilian and the other women were in constant danger, and overworked. Air raids were common as German bombers searched for Serbian infantry formations in the mountains.

The dressing station comprised 40 beds under canvas tents and a makeshift operating theatre in a shed. The patients slept on modified pallets made of straw and blankets stretched over iron frames to elevate them off the ground. For sicker patients the beds were elevated on empty shell cases to make tending casualties less back-breaking.

Here, 5,000 feet up Mt Kajmakcalan, overlooking the minarets of Monastir and the snow-clad mountains behind, Dr Cooper performed operations day after day in the harshest of conditions. Clearly happy here, Lilian saw this as the culmination of her life's work: she was performing real surgery, and making a real difference. Despite the strain of long hours, she performed multiple amputations, removed shell fragments, bullets and other shrapnel from men's bodies ravaged by war. Through life- and limb-saving surgery, patients were stabilised and then transported by ambulance to the main hospital at Ostrovo. Countless lives were saved in this way.

From mid-January to the beginning of March – a period of six weeks – the ambulances brought down 1,840 patients from the dressing station. Over the next seven weeks a further 523 casualties were admitted, of

---

[73] Williams, L. M. p. 34.

whom 60 died from their wounds. During her eight-month period at the 40-bed dressing station, only 16 of the 144 patients Dr Cooper operated on died.[74]

At Dobraveni, there were more than just the effects of combat to be worried about. Spring brought warmth and relief from the bitter cold, but also flies. Flies breed in manure, human waste and decaying organic matter, transmitting dysentery, cholera and typhoid. Dobraveni was overrun with flies and flooding. The kitchens were infested and the environment was ripe for illness.

Typhoid, much feared for its ravaging and indiscriminate effects, broke out within the nearby Russian detachment allied with the Serbs. Lilian dealt with this outbreak in her usual, capable way. The Russian commanding officer is said to have been so impressed with her skill and commitment that she was awarded the Imperial Russian Order of St Anne. No record of this award exists, however, even among her possessions; although it is mentioned in one of Agnes Bennett's letters home.[75]

The conditions eventually took their toll. Not surprisingly given her age and the appalling conditions she had to tolerate, Lilian became gravely ill with bronchitis. Eventually, she had to leave her post and convalesce in Salonika.

Her year of service now complete, she decided that she was no longer well enough to continue. She was replaced by another Australian woman from Victoria, Dr Mary de Garis.

Although the service of Dr Cooper and her team were not recognised in Australia, it was not overlooked by Britain's allies, the Serbs. In March 1917, Dr Cooper, Dr Bennett and Miss Bedford were each awarded the Order of St Sava – Serbia's medal for humanitarian service. They received their awards personally from the Aide-de-Camp to Serbia's Crown Prince Alexander Karageorge.

The citation (Order No. 2218 of the Commandant of the Third Serbian Army dated 12th March 1917) reads:

---

[74] ibid.

[75] This is recorded in De Vries *Heroic Australian Women in War* on p. 90. However, it is likely that either it did not occur or that Lilian did not value the award as many were handed out during the campaign.

The Great War: an empire at war

> The doctors of the 3rd Surgical Field Hospital [including] Dr Lillian Violet Cooper, ... have shown ... very zealous and very diligent care of our most seriously wounded soldiers, carrying out their work with extraordinary will and a real love for our fighting men. In the name of all the officers, non-commissioned officers and soldiers of my Army, I thank them for all the services they have rendered to my soldiers.
>
> Signed Commandant, General Vashish[76]

Dr Jessie Scott, of New Zealand was also included in the commendation and received the Order of St Sava. Miss Bedford was recognized separately in Order No. 2219 on the same date:

> The Chief of the Automobile Section of the 3rd Surgical Field Hospital, Josephine Bedford, with great diligence and zeal, has successfully directed the work of evacuating our wounded. In her work, she has shown herself diligent and untiring and very often she has personally taken part in the evacuation of our wounded. At all times she has distinguished herself by her sympathy and her love for our soldiers and for this I thank her with all my heart.
>
> Signed Commandant, General Vashish

Back at the main hospital in Ostrovo, spring brought its own problems: sickness became rife. In war, disease is as much an enemy as bombs and bullets, and this campaign was no exception.

As well as flies, the warm weather brought mosquitoes, breeding in the marshy swamps near Lake Ostrovo. All the staff took doses of quinine and mosquito nets were compulsory. Despite these measures however, Sister Agnes Kerr became gravely ill with malaria and died. Miles Franklin was working as a camp hand, predominantly in the kitchens, and she too became severely ill with malaria.[77]

Malaria, with its characteristic quartian fever, can be ignored for a

---

[76] Archival Files of the Scottish Women's Hospitals for Home and Foreign Service, The Mitchell Library, Glasgow, Scotland, Tin 36, quoted by Williams in *Behind the Front Lines in Serbia*.
[77] De Vries, Susanna, *The Complete Book of Great Australian Women*, Harper Collins, 2003, p. 431–506.

while.[78] For one or two days there is a reprieve, but with precision and certainty the fevers return at regular intervals. Sweats, chills, joint pains and headaches progressively take their toll. Soon, without effective doses of quinine, the complications of malaria set in – respiratory illness, fits, coma and ultimately death.

Despite her constitution, Dr Bennett was not spared. She tried to ignore her illness, but became progressively weaker. As the direction of the war changed a decision was made to close down America Unit No. 1. Plans were put in place to return the staff to Britain and a small skeleton-staff were left to look after the patients too ill to move.

By now, Dr Bennett was well in the grip of malaria. She was weak and collapsing regularly. Just before leaving the Ostrovo site the hospital was hit by Austrian shelling. Patients and staff alike were wounded and the operating theatre set ablaze. Dr Bennett was the last Australian to leave before the hospital was overrun by Bulgarians. They killed all remaining staff and patients. No-one was spared in the massacre that followed.

Dr Bennett returned to Britain and worked at a British military hospital at Netley in Southampton until the end of the war.[79] For her efforts in Serbia she was awarded the Order of St. Sava 3rd class and the medal of the Red Cross of Serbia. On her way back to Australia she worked as a doctor on one of the troop ships – caring for wounded ANZACs who made it back to the sunburnt country they had left in innocence just a few years before.

SERVING WITH THE BRITISH

By 1916, the British Army saw a need to include women doctors in its medical teams, but it was unclear how.

Even though women were formally recruited into the RAMC from April 1916, their position and entitlements within the army's organisational structure remained unclear. As a result, anomalies and inconsistencies were the norm, affecting women's status, rank and employment conditions.

The Medical Women's Federations made repeated petitions to remedy

---

[78] Quartian is commonly used in the medical profession to mean once every four days.
[79] De Vries, *Heroic Australian Women in War*, p. 93.

the situation and there were appeals to Churchill himself. Still, their calls for equality fell on deaf ears. The RAMC's women doctors wore rank and drew pay from the War Office. Yet, their rank was never formally recognised – or gazetted – and they were not entitled to veterans' allowances after the war. Their relegation to third class travel on trains – while male officers of equal rank travelled in greater comfort – was an issue that some women found particularly galling.

One British doctor, Edith Guest, summed up the feelings of many women doctors in a letter home in 1918, after serving in Malta and Egypt:

> … we are exactly as we were when we first joined, and although we are senior in service to many of the men here, yet they all – however young and inexperienced – rank above us, and any youngster will take precedence of us even if we serve ten years. The longer one serves, the more galling this becomes.[80]

Discrimination against women was also evident in the honours and awards system. Here, the war service of one Australian doctor in particular – Phoebe Chapple – is a case in point. She became the first Australian woman and the only female doctor to be awarded a Military Medal for gallantry.

Major Phoebe Chapple MM

Phoebe Chapple entered the University of Adelaide aged just 16. In 1904, she graduated in medicine amongst a group of pioneer women doctors. Born to a privileged family, her father was headmaster of Prince Alfred College, an elite Adelaide boys' school, and supported her desire for an education. After she qualified as a doctor Phoebe would sometimes travel in a phaeton driven by a liveried coachman to visit her patients.[81] But her

---

[80] Dr Turnbull to Dr Sanderman, 14 November 1918, CMAC:SA/MWF/C.163; Dr Guest to Dr Walers, no date [c. February 1918] CMAC:SA/MWF/159 cited in *Medical Women at War*, p. 172.

[81] Blanche, Craig, 'Dr Phoebe Chapple: The first woman doctor to win the Military Medal', Canberra, Australian War Memorial online, 2009 Available at: http://www.awm.gov.au/blog/2009/06/30/dr-phoebe-chapple-the-first-woman-doctor-to-recieve-the-military-medal/

early career also showed an attitude of service; and she frequently treated the poor and disadvantaged, working with both the South Australian [Women's] Refuge and the Sydney Medical Mission.

When war broke out in 1914, Phoebe followed events with interest. There were calls for volunteers; there were appeals for doctors at the front. Phoebe wanted to contribute, but as the Australian Army refused to appoint women as doctors, Phoebe's only option was to serve with the Allies, and go to Britain.

Frustrated with the Australian Army's refusal to appoint women doctors, she travelled to England in February 1917 to enlist in the RAMC. The RAMC appointed her to the Cambridge Military Hospital in Aldershot, where she was in charge of one of the surgical wards.[82] The wards were busy, with convoys of wounded arriving continually from France. Phoebe's work soon gained recognition and she was the first woman surgeon in the hospital to receive equal status to the men. She was given the rank of "Major", although the British Army did not gazette her rank formally. This situation was common to all of the women doctors and would have significant consequences for recognition of their service in years to come.

Later, she was attached to Queen Mary's Army Auxiliary Corps (QMAAC). As members of the QMAAC, women doctors examined the recruits and were in charge of the health of women serving. Their uniforms were nurses' uniforms and many found this galling, resenting the fact that they were not formally recognised – or gazetted – as working with the RAMC.

Dr Chapple was one of the first two women doctors sent to the front, which she "regarded as an honor [sic] for Australia".[83] By this stage Britain was struggling to provide enough medical manpower at the Front and this opened the door to women who wanted to serve in Europe. It was never intended that women should come under fire, but the boundaries soon became blurred.

In November 1917, Phoebe left for France. Here she was attached to

---

Accessed 10 May 2013.

[82] The *Register*, (Adelaide, SA), 3 September 1919, p. 7.

[83] See: Blanche.

the Women's Auxiliary Army Corps at Abbeville, and found herself – to use her own words – "in the centre of the battle zone".[84] It was her actions six months later however, that would see her become the first (and only) female doctor to be awarded a Military Medal for gallantry.

On 29 May 1918, Dr Chapple was inspecting the QMAAC Camp 1 near Abbeville when it came under a German aerial bombing attack. QMAAC Camp 1 accommodated women serving with the Queen Mary's Army Auxiliary Corps, who were working at the British hospital in Abbeville, quite close to the Australian military hospital, known as 3AGH.

Earlier that day German planes had been seen overhead. Dr Chapple thought they were taking photographs. She did not know whether they had spotted the hospitals on site, but the area was certainly subject to regular bombing raids.

The details of what happened next remain sketchy. According to one historian, a lorry was set on fire close to the camp and "by the light of the flames German aircrew were able to drop three bombs on the compound".[85] Two bombs destroyed huts while a third exploded on a covered trench used by the women as a shelter. It is unclear whether the lorry was set on fire deliberately, or was hit in an earlier raid. What is clear is that the impact of the three bombs was devastating: one hit a trench where the women were sheltering.

Out of 40 women, eight were killed outright and eight others injured, one dying later of wounds. Hampered by darkness and difficulty, subject to yet more raids and with communication with headquarters temporarily cut off, Dr Chapple worked her way through the trench, tending to the wounded. Finally, at two o'clock in the morning, the administrator in charge of the section called out the roll. A QMAAC historian wrote:

> No-one who was there will ever forget the silence that was only broken by a little gasping sob from someone when a name was called and not answered.[86]

---

[84] The *Register*, (Adelaide, SA), 3 September 1919, p. 7.
[85] Cowper, Julia, *A Short History of Queen Mary's Army Auxiliary Corps* (London: Women's Royal Army Corps Association, 1966), p. 53.
[86] ibid.

Phoebe was one of the fortunate. She had survived.

The citation for her Military Medal reads: "For gallantry and devotion to duty during an enemy air raid. While the raid was in progress Doctor Chapple attended to the needs of the wounded regardless of her own safety."

While Dr Chapple sought to play down her involvement, others became concerned, and even angry at the discrimination they felt she received. She should have been awarded a Military Cross, in their view, which was more usual for an officer. A colleague of Dr Chapple's from Adelaide, Dr Helen Mayo, was so incensed at the discrimination surrounding awards that she noted, many years later: "Had [Chapple] been an officer (and a man) she would have received the Military Cross".[87]

The issue of Military Medals for female gallantry was contentious in the Great War. Some felt that the Military Cross (MC) was more appropriate, but it was not sanctioned. Women did not hold commissions and were never formally gazetted for their roles or actions.

Awarding the Distinguished Service Medal to women was also ruled out. There was concern that this would offend some of those who already held it. Instead, the relatively new Military Medal (MM) – instituted in 1916 – was agreed upon, and by Royal Warrant it was extended to include women who displayed 'bravery and devotion under fire'.[88]

As a result, the MM could 'under exceptional circumstances, on the special recommendation of a Commander-in-Chief in the Field, be awarded to women'.[89] No restriction was made on the grounds of nationality or service. Some considered an MBE (Member of the Most Excellent Order of the British Empire) was more appropriate for women, but Dr Chapple was awarded the MM as it was the only award that could be made to a woman for gallantry in the field.[90]

---

[87] See: Blanche.

[88] Stewart, Elizabeth, 'Nurses under fire', *Wartime*, (Canberra: Australian War Memorial, 2010), http://www.awm.gov.au/wartime/50/stewart_nurse/, Accessed 10 May 2013.

[89] See: Blanche.

[90] According to the Australian War Memorial, 9,926 Military Medals were awarded to AIF troops during World War I, not including the awarding of bars. Women represent just 0.1% of recipients of the medal throughout its history: Chapple became a member of an exclusive group.

Undeterred by her experiences at the front, Dr Chapple went on to serve as a doctor with the British Army in Rouen and Le Havre until the end of the war. She returned to Adelaide in 1919 and immediately resumed practice, continuing to work until she was 85. In 1953, she was one of only a few Australians invited to the Queen's Coronation Ceremony held at Westminster Abbey. She also attended the Medical Women's International Association Conference in Edinburgh and took the opportunity to visit German and European medical centres.

A tall and proud woman, she marched each year on Anzac Day, ironically at the head of the nursing units. Although her life and work were largely defined by her wartime service, she preferred to play down her experiences.

Phoebe Chapple died on 24 March 1967 and was cremated with full military honours.

Women in the RAMC

Other Australian women doctors were equally eager to serve in their professional capacities and when positions opened served with the RAMC. These included Doctors Elaine Little, Eleanor Bourne and Katie Ardill.

Captain Elaine Little

Elaine Little graduated from Sydney University in 1915, securing a residency post at the Royal Prince Alfred Hospital. Although she was employed to replace the hospital's pathologist who was away at war, Dr Little was also keen to contribute to the war effort. Like other women before her, Dr Little was turned away by the Australian Army Medical Corps. Also like others, she paid her own way to England and approached the RAMC. She was accepted into the Corps in 1918.

Initially, she worked at the Lister Institute of Preventive Medicine in London, but was soon called to France. There she worked as a pathologist in the 25th Stationary Hospital and subsequently the isolation unit (46th

Stationary Hospital) at Etaples.[91] Dr Little often worked late into the night; performing bacteriology, serology, vaccine preparation and blood tests. Her work came to prominence in 1918 when an epidemic of influenza caused almost as many deaths as the battlefield. The disease swept across Europe, destroying bodies worn down by constant fighting, poor food supplies and – as for those who had seen action at the front – the squalor of the trenches.

Captain Katie Ardill (-Brice) OBE DStJ

Dr Katie Ardill was also an Australian doctor who served with the Royal Army Medical Corps. She graduated in medicine from the University of Sydney in April 1913. In 1914 she joined the St John Ambulance Brigade's Nursing Division. In 1915 she became Divisional Surgeon, the first female doctor in New South Wales to hold the post.[92]

At the outbreak of the Great War, her services too were rejected by the Australian Army.[93] Dr Ardill duly made her own way to Egypt and on to England, where her initial appointment was to the County Middlesex War Hospital in St Albans.[94] Previously, the facility had been used as the County Asylum and re-opened in September 1915 to receive wounded soldiers from the Western Front.

The hospital remained open until late 1919. It was a large hospital with a total of 1,600 beds. It also had a specialist military mental unit with capacity for 250 men – recognizing the mental health injures sustained of soldiers returning from the front.[95] Both facilities were staffed and operated by the RAMC and Queen Alexandra's Imperial Military Nursing Service,

---

[91] Heagney, Brenda, "Little, Elaine Marjory (188401974)'. *Australian Dictionary of Biography*, National Centre of Biography, Australian National University, http://adb.anu.edu.au/biography/little-elaine-marjory-10838/text19231. Accessed 27 April 2012.

[92] Heather Radi, 'Ardill, Katie Louisa (1886–1955)', *Australian Dictionary of Biography*, National Centre of Biography, Australian National University, http://adb.anu.edu.au/biography/ardill-katie-louisa-5624/text8413, Accessed 13 June 2013.

[93] ibid.

[94] 'Dr Katie Ardill-Brice (née Ardill)', wikispace entry, Available at: http://stjohnambulancensw.wikispaces.com Accessed 16 June 2014

[95] See http://www.1914–1918.net/hospitals_uk.htm

supplemented by voluntary workers from a number of organisations including the Voluntary Aid Detachments, the Red Cross, and St John's Ambulance. It is unclear in what capacity Dr Ardill was engaged here.

Dr Ardill subsequently served at the Anglo-Belgian Base Clearing Military Hospital at Étaples in France. Here she held the rank of Captain, and her service is recorded in the Record of Special Reserve Officers' Services.[96] She served for four years in total, including a stint in Egypt from May 1918 until June 1919.

Others that had already seen service with the Scottish Women's Hospitals also now joined RAMC units. Elise Dalyell used her experience with the RAMC in Malta and Salonika. Agnes Bennett also returned to duty with the RAMC after recovering from malaria and her time in Serbia.

PHYSIOTHERAPISTS IN THE GREAT WAR

Australia's women doctors were not alone in experiencing rejection and frustration. Physiotherapists also fought bureaucracy and struggled for due recognition.

To their advantage, the Australian Army considered the professional skills of female physiotherapists to be essential and their skills were integrated into Australia's medical services. Still, their journey to acceptance was not without ambiguities and their contribution has been underrated.

During World War I, Australian women were employed as masseuses within the Australian Imperial Force (AIF) and served in England, France and Egypt. At first, the Australian Surgeon General, Sir Neville Howse VC, an esteemed veteran of the Boer War campaign, deemed masseurs and masseuses as 'unnecessary' to the AIF.[97] In his view, massage could be provided equally well by others such as nurses and orderlies and there was no need to invest in physiotherapy equipment. Howse also opposed

---

[96] Record of Special Reserve Officers' Services, Army Book No. 82, Entry 118, held at the Army Medical Museum, UK.

[97] Neville Howse was Australia's first Victoria Cross winner, awarded for his actions at Vredefort during the Boer War in 1900. He remains the only Australian member of the military medical services to have received such an honour.

commissions for masseurs and masseuses, considering their services to be a 'waste of money as far as the AIF was concerned'.[98] His view was challenged by the so-called 'Six Months Policy'. This dictated that soldiers needed to be fit enough to be sent back to the army within six months of starting treatment or be transferred back to Australia. Most of the wounded from the Gallipoli campaign were sent to Cairo via the island of Lemnos, just west of the peninsula, which housed two Australian stationary hospitals. From Cairo, depending on the severity of their injuries, soldiers were either returned to Australia on hospital ships or opportunity transports. Hospital ships were increasingly filled with casualties of war with major injuries in need of complex splints and continued treatment, often during many weeks at sea. There was recognition that more soldiers could be returned to the front if they were successfully treated and rehabilitated in Egypt.[99]

As a result, in August 1915 authority was given for each of the Australian Military Districts to establish an Army Massage Service. In this regard, Australia was ahead of the British service, which had yet to include masseurs and masseuses on its payroll. Instead, the British relied on private benevolence to fund the employment of massage services, which were restricted to hospitals and convalescent depots. The British War Office did not operate a massage service outside of Britain until as late as 1919.

Legally, like their nursing colleagues, the masseuses did not hold military rank. To add to the confusion they were appointed as 'Staff Nurses', and were addressed as 'Miss' rather than 'Lieutenant' or 'Captain'. Their chain of authority was ambiguous, and this ambiguity was carried forward to subsequent wars. They took their orders from the medical officer, but for all other matters they came under the command of the matron. Male masseurs, on the other hand, received more equitable treatment within the AAMC.

The uniform of the masseuses was also modelled on that of the nursing corps. No flesh was shown with long sleeves and a Rising Sun badge attached to a high collar. A full-length white apron (to be changed every

---

[98] Butler, A. G., *Official History of the Australian Army Medical Services, 1914–1918*, 1940.
[99] ibid.

day) was worn over a dress and the outfit was topped with a fine linen cap (allowing no hair to be shown).[100]

Six masseurs (appointed with the pay and privileges of staff sergeants) and 12 masseuses (appointed with the pay and privileges of staff nurses) embarked with AIF forces headed for Egypt and England. Among the first to leave were Mrs Victoria Shorney and Miss Edyth Cocks. Each was presented with an eiderdown quilt before their departure for service in Egypt – an appropriate gift given the number of patients presenting to the Australian military hospital in Cairo with frostbite.

The work of the masseuses involved more than mere massage – they provided hot-air baths and thermal treatment using mild electric shock (faradic) stimulation and heat lamps. Open wounds were treated with zinc or copper ionization, joints were mobilised and disfiguring scars from burns were manually softened and stretched. For those suffering the effects of frostbite, massage had to be performed carefully to minimise the agonising pain that accompanied the loss of toes.

By December 1915, one masseuse was allocated to each hospital ship to assist with rehabilitation and the management of wounds. Later, they were included on transport ships back to Australia to ensure the continuity of patients' treatment. It would not be until the next world war however, that the real benefits of the emerging practice of physiotherapy were fully appreciated.

Between the Great Wars

The medical women of the Great War had shown dedication and resilience despite bureaucracy, opposition and countless instances of discrimination. They had proved themselves equal, demonstrating:

> [There] was really nothing that a woman doctor could not do in a war zone. They treated virtually every kind of wound and disease, they underwent the same hardships, privations and dangers as men, [became prisoners of war], took part in

---

[100] Wilson H., *Physiotherapists in War. The story of South Australian Physiotherapists during World Wars I and II, Japan–Korea and Vietnam*, Gillingham Printers, Adelaide, 1995, p. 2.

devastating retreats, and worked under shells and bombs. None of this valuable experience advanced their career prospects. And it took the outbreak of another world war to gain medical women commissioned rank in the [Australian] Army.[101]

In Britain, once women had the vote, the suffrage leagues were disbanded. Arguments that women, if employed as doctors with the armed forces, should receive temporary honorary commissions continued to fall on deaf ears. The war was over and they had the vote; the role of women was no longer considered a pressing issue.

In C. E. W. Bean's official history of the Great War there is little mention of the Australian women who served in non-nursing military roles. There is also no mention of the women who were awarded Military Medals or Medals of Gallantry for their service as doctors during the war, or of the women who, having been rejected by the AIF, decided to travel to Europe at their own expense and served, with distinction, with Australia's allies in Britain, France, Belgium, Malta, Egypt and the Western Front.

There is no mention of their involvement or contribution to military medicine. There is silence about their service in treating Allied, British and Australian casualties of war. Despite significant foreign awards from the governments of Britain, France, Serbia and Greece, their service has passed largely unnoticed in their own country – Australia.

C. E. W. Bean's omission helped to reinforce the role of women during the Great War as 'carers' and 'nuturers'. This stereotype, that was widely held in Australian society during the postwar period, did not sit comfortably with those women who were more adventurous, independent and determined to pursue professional medical roles.

In subsequent histories of the RAAMC, the service of these women has also been overlooked. This, in part, is due to technicalities. In some cases they were recorded as 'attached' to Australia's medical corps during the Great War and did not receive formal commissions; in other cases their service was with Australia's allies.

---

[101] Leneman, p. 177.

Homecoming

Little is known of the women's homecoming. Perhaps overshadowed by the sheer numbers of 'returned' servicemen they simply reintegrated into the uneasy peace of a society, coming to terms with the legacy and wounds of war.

Australia wanted to return to 'normalcy' as soon as possible. The Great War had been seen as a 'great struggle for humanity' and the first in an era of 'total war' where every member of the community was in some way affected. With the Armistice came hope. Perhaps the war had been an aberration – there was a desire to return as quickly as possible to 'pre-war' values and lifestyles.

Women in Australia rapidly resumed more traditional roles – they became wives, daughters and carers. Women from more privileged backgrounds were expected to return to the drawing rooms or tennis court as if the war had never happened. There was pressure to forget the war and move on. Any hopes or expectations that Australia's medical women could pursue their professional roles within a military context, or use their service to boost their careers, were soon dashed. While service in 'The Great War' boosted the status of male professionals, the same could not be said for women.

Following demobilisation there was no role for women in the AIF. Those who had served returned to civilian practice and their careers reflected the options open to them at the time. For women doctors, it was a case of finding jobs in 'acceptable' positions in women's hospitals, asylums or maternal health and childcare.

Physiotherapists fared marginally better. During the 1914–18 war there had been delays in their employment with military units, but once deployed their importance was quickly established and there were efforts to ensure that a 'standing massage unit' would be available for future conflicts. By 1938, the Australian Army allocated five masseuses per 600-bed hospital and 10 per 1,200-bed hospital.

What, then, was the legacy and effect of their wartime experience for Australia's female pioneers of military medicine? War demonstrated that they could cope in stressful conditions that were at least as demanding as those faced by men, and often at great personal risk. They proved that women could run large organisations efficiently, manage epidemics of disease and command both respect and discipline.

For some women, their first exposure to the realities of war had been harrowing. They had to learn quickly, overcoming their lack of experience in dealing with battlefield surgery and complex injuries. Their work with wounded men, often in appalling working conditions, proved their ability beyond doubt. It also gave them confidence in their own intellectual, practical and administrative capabilities.

Their experience of war undoubtedly made the women represented here more resilient and more confident in their clinical skills. Some continued to demonstrate the leadership and pioneering spirit that characterized their service during the war years, devoting themselves to issues of social reform, women's rights and health issues. They asserted their professionalism and independence in new ways, through emerging movements, campaigns and organisations.

Vera Scantlebury (formerly of Endell Street, see above) used her newfound self-confidence to become a pioneer in the field of infant welfare in Victoria. She became the first woman to be appointed head of a government department in that State.[102] Her time at Endell Street hospital had not only given her self-confidence; it gave her an understanding of large-scale logistical management and planning. Above all, perhaps, she realised what it was possible for a woman of her experience to achieve. She was heavily influenced by Major Louisa Garrett Anderson and Dr Flora Murray as role models.

Elsie Dalyell returned to the Lister Institute in London initially, before embarking on her most important contribution to medical science. Together with Dr (later Dame) Harriet Chick, Elsie was part of a small team that travelled to war-torn Vienna. Over the next two years she demonstrated conclusively that rickets (then a scourge of poverty-torn Europe) was due to a deficiency of Vitamin D and could be prevented with cod-liver oil and exposure to sunlight.[103]

---

[102] See: Sheard.

[103] Chick, Harriet; Hume, Margaret and McFarlane, Marjorie, *War and Disease: A History of the*

When Dalyell returned to Australia, however, she was given work that was far more mundane. Despite her experience and war record she had difficulty obtaining a suitable appointment in Sydney and her war service had done little to advance her professional opportunities. She went on to become an acknowledged expert in syphilis testing and spent the rest of her career specializing in venereal disease (VD): "… an inadequate use of a splendid mind and a forceful and most engaging personality."[104]

A number of Australia's medical women ended up running VD clinics on their return home. In 1916, the disease was rampant among Australian troops, particularly in France where prostitutes were readily available. When they returned home they brought their diseases – as well as their experiences – with them. One legacy of the Great War was an epidemic of VD that was spread to Australian wives and girlfriends, all before the introduction of penicillin to counter the disease.

Lilian Cooper went on to become a Foundation Fellow of the Royal Australasian College of Surgeons and was its first female Fellow (admitted as number 128 on 17th June 1927). This was the supreme accolade for a woman who had worked hard to be accepted as a surgeon. She was elected by her colleagues as one of the top 200 surgeons in the country. Her stationery marked her achievement: she became Lilian Violet Cooper MD, FRACS.

Dr Cooper remained unconventional, strident and outspoken throughout her career. She earnt the professional respect of her colleagues, and as a tribute to her, her companion, Miss Bedford, bequeathed the land where the Mt Olivet Hospital in Brisbane stands today. Bedford also donated a set of double Gothic stained glass windows – the soldier windows – that grace St Mary's Church at Kangaroo Point, nearby.[105] Cooper died in 1947 and is buried with Miss Bedford in Brisbane.

After her return to Australia in 1919, Katie Ardill married Charles Brice in 1921. In conjunction with the Returned Services League, she established free medical clinics in Western Sydney to treat the wives and

---

Lister Institute, Andre Deutch, 1971, Chapter 15.

[104] Richardson, G. D., 'The Dalyells and their Kin', *quoted in* Crofton, Eileen, *The Women of Royaumont*, p. 269.

[105] Crammond, T., 'Lilian Violet Cooper MD, FRACS, Foundation Fellow, Royal Australasian College of Surgeons. *Australian and New Zealand Journal of Surgery*, 1993, vol. 63, pp. 134–142.

families of ex-servicemen. She also renewed her association with St John, becoming the first woman to chair the association in 1949.

Her military associations continued and the lessons of war clearly influenced her role within St John. In 1934 she was chosen to represent the organisation at the opening of the Anzac Memorial in Sydney by Prince Henry, Duke of Gloucester. She gave addresses on the work of St John in war and peace and participated in training events, including stretcher drills. In 1952, Dr Brice (née Ardill) attended the British Military School of Civil Defence and studied a course in atomic bomb defence.[106] In 1942 her services to medicine were recognised with the award of an OBE.

Agnes Bennett also continued to pursue her career. Hoping that finally her experience would be recognised, she tried again to find a position as a surgeon in Sydney, but with no more success than she had before the war. Accepting rejection in her own country, she returned to New Zealand and set up practice. She became the first woman in New Zealand to become superintendent of a public hospital.

Bennett's sense of adventure and desire to challenge herself professionally remained undiminished. She spent time working in the remote Chatham Islands and with the Royal Flying Doctor Service in Burketown in North Queensland. Later, after her services had been rejected again in World War II (this time by the New Zealand Army) she became a lecturer, and also undertook a medical rescue mission to Antarctica – an environment even more hostile than the mountains of Serbia. She was awarded an OBE for her services to medicine.

One of the few women to be formally recognised for her service by Australia was Millicent Armstrong, recipient of the Croix de Guerre. On her return home after the war Armstrong made an application for land under the Returned Soldiers Settlement Act in 1921. She was granted a title to 1028 acres (416 ha) with a capital value of over £2600 at Gunning, 75km north of Australia's new capital – Canberra.

Armstrong's experience was unusual, however. Although their achievements were known to their families and friends, Australia's medical

---

[106] See: 'Dr Katie Ardill-Brice (née Ardill)', available at: http://stjohnambulancensw.wikispaces.com Accessed 15 November 2012.

women of the Great War remained largely unrecognised by the country they considered home. It was not until another war descended on Europe that the contribution of Australian women to military medicine was revisited, and their place in history restored.

CHAPTER THREE

World War II: a corps in transition

Today we are sending youized out to battle. Only you are not armed with poison gas and bayonets, but with antiseptic, chloroform and healing hands.

Budd Schulberg, U.S. screenwriter and intelligence analyst, speaking at the Writers' Radio Theatre 1941.

Vital accomplishment and not sex should be the measuring rod.

Dr Emily Dunning Barringer, President of the American Medical Women's Association, 1942.[1]

On 3 September 1939, the sombre words of Australian Prime Minister Robert Menzies were broadcast to the nation:

> Fellow Australians, it is my melancholy duty to inform you officially, that in consequence of a persistence by Germany in her invasion of Poland, Great Britain has declared war upon her and that, as a result, Australia is also at war. No harder task can fall to the lot of a democratic leader than to make such an announcement.[2]

With those words, any lingering delusions that the fragile peace would hold were singularly extinguished. Australia was again a country at war.

---

[1] Bellafaire & Graf, *Women Doctors in War*, 2009 Texas A&M University Press p. 62.

[2] See: http://www.ww2australia.gov.au/wardeclared/

Once again, Australian military forces were mobilised in the interests of the Empire and the nation and, once again, Australian medical women would play their part.

World War II would be a period when medical women found greater professional legitimacy in their army roles. Paradoxically they participated less overseas, undertaking more significant 'home guard' roles, establishing a foundation for the future role of women in the Royal Australian Army Medical Corps (the RAAMC, also referred to here as 'the Corps').[3]

Exclusion through fragmentation

At the outbreak of World War II, Australia's Navy and Air Force employed medical women within consolidated women's services. The Australian Army was different, however. It established three separate services for women during the war.

The first was the Australian Army Nursing Service (AANS). Established in 1902, AANS amalgamated various nursing services from the colonial era and formed part of the Australian Army Medical Corps during the Great War. In World War II, the service was mobilised again, and after the war, in 1948, it was renamed the Royal Australian Army Nursing Service (RAANS). In 1942, the Australian Women's Army Service (AWAS) was formed, followed by the Australian Army Medical Women's Service (AAMWS) in December 1943.

The roles of each service differed somewhat and the result was a fragmented system where the contributions of professional medical women were either excluded or overlooked. Women experienced the same vagaries of service as their World War I forebears. There were discrepancies in pay, rank, entitlement, uniforms and status. It was common for professional medical women to serve with multiple agencies throughout the war, leaving a complex history of employment where individual contributions are hard to appreciate.

Consequently, the service of medical women in the Australian Army

---

[3] The 'Royal' prefix was added to the name of the Australian Army Medical Corps (AAMC) in 1948; the AAMC was formed in 1902, post-Federation.

of World War II remains largely unrecognised. In Patsy Adams-Smith's *Australian Women at War* less than a single page is dedicated to "Persons and Particulars": the female doctors, scientists and other medical professionals who served their country with distinction during World War II. It was a conflict that saw pioneering achievements on the part of women; landmarks in the history of the Corps.

WOMEN'S STATUS AS DOCTORS

In the years between World War I and World War II, the professional status and prospects of medical women had advanced considerably. By the end of the 1930s more hospital residency posts were available for doctors; medical research and scientific positions had expanded to include women graduates. Women now had unprecedented opportunities to work away from home, or, if young and single, to travel and seek adventure.

And yet, paradoxically, the Australian Army did not reflect these changes.

Despite the increasing number of women graduating from medical schools, women undertook very different operational roles during World War II than in World War I. There is no evidence that they took the same measures to travel overseas or set up their own military hospitals independently, as they did in World War I. Indeed, the only Australian women who would serve as army doctors overseas during World War II – after, again, being rejected by their own country – were those who wore the uniform of their British allies.

The limited involvement of women doctors in the Australian Army stands in stark contrast to their inclusion in Britain. During World War II, 600 women doctors served with the British Army. Whilst they provided limited service – "… we may not serve on the front line"[4] – they still found their way to operational theatres of war. Sixty women doctors served as specialists in 14 different areas of expertise; all the larger British General Hospitals in France and north-western Europe included women among their medical staff.

---

[4] Wimmer, Albertina, *The British Woman Army Doctor*, unpublished manuscript courtesy of the Royal Army Medical Corps Museum, Ashvale, UK p. 6.

World War II: a corps in transition

In the United States, American women doctors also struggled to carve out a legitimate role within the Army. Entering the war late, they were restricted to service within the Medical Reserve Corps, with roles at home or limited to 'contract positions' in England attached to the Royal Army Medical Corps. Only a select few were assigned to positions in Europe or the Pacific theatres of conflict. Like their Australian counterparts, American women doctors suffered similar vagaries of corps affiliation, with most serving under the Women's Army Auxiliary Corps (WAAC) or the Women's Army Corps (WAC). Despite acknowledged shortages of over 2,000 doctors, the U.S. persisted in its refusal to commission medical women.[5] There was vocal opposition to this policy and Dr Emily Burring Barringer, President of the American Medical Women's Association, argued that since:

> ... women had to meet the same standard as male physicians to practise medicine, they should be considered equally qualified when it came to serving their country.[6]

There was no such outcry in Australia. Perhaps, in late 1930s Australia, medical women were less radicalised by the suffrage agenda of their forebears or felt more secure in their professional positions. There was little public protest about a lack of opportunity. Instead, pragmatism won out and they took on roles in Australian hospitals left vacant by their male counterparts who had gone away to war. If a woman's place was not in the home, at least it could be on the home front.

Perhaps too, the horrors of World War I had permeated Australian society and there was not the same naïve enthusiasm or sense of adventure that existed in 1914. Few Australian families had been spared the lasting effects of World War I – they counted relatives and loved ones amongst the dead; they lived with the wounded and the psychologically scarred.

Australian medical women of the early 1940s were now wives and mothers with established lifestyles and commitments. It was largely felt

---

[5] Bellafaire & Graf p. 62.
[6] ibid.

that their services would be best used at home in a civilian capacity, freeing up male medical manpower to enlist in the Corps:

> The attitude to women and work had been transformed by war. It had become clear that every fit man was needed in the front line and that every fit woman who could do a man's job ... must be mobilised to do so.[7]

Planning for full mobilisation revealed that 1,160 medical officers would be required for the Australian Army Medical Corps (the predecessor of the RAAMC) with reinforcements required at the rate of 10 per cent per year. Australia's official medical historian of World War II, Dr Allan Walker, noted:

> In Australia [at the outbreak of the Second World War in 1939] [sic] detachments of both men and women were trained in first aid and in the elements of hygiene and home nursing by the Order of St John, the Australian Red Cross Society or an approved Ambulance Association. No other provision was then made for women in the Services... [but] the suggestion was made at that time that physiotherapists and certain other specialists might be regarded as a special variety of voluntary aid ...[8]

Despite the experiences, commitment and achievements of women doctors during World War I, little had changed in terms of their mobilization. In planning Australia's contribution to the war effort, the same attitudes and exclusion prevailed. For example, unlike nurses and physiotherapists, women doctors were not included in Australian Army Reserve units.[9]

When World War II broke out, there were 323 female and 5,083 male doctors registered to practise in Australia. In 1941, the Deputy Chairman of the Medical Co-ordination Committee advising the Australian

---

[7] Wimmer, 6.

[8] Walker, Allan S., *Australia in the War of 1939–45: Medical, Middle East and Far East*, Canberra, Australian War Memorial, Canberra, 1953.

[9] ibid.

World War II: a corps in transition

Government, Sir Alan Newton, submitted a summary of the medical manpower available for the war effort to the Secretary of the Department of Defence.[10] His analysis may have overestimated the number of doctors registered; it also excluded women:

Total medical practitioners in Australia	**6,500**
Deducted –	
Women doctors	500[11]
Males over 60	500
Medically unfit	254
Essential Govt services	500
Total	2,424
[Number remaining for War Service]	**4,076**

Some senior officers were not adverse to the idea of women doctors serving in the Corps, but this was by no means a universal attitude. Major General Frederick Maguire (Director-General of Australian Army Medical Services 1941–2) is quoted as saying: "The more they are used for certain types of work the better."[12] The meaning of 'certain' remains unclear.

By early 1940 it was finally accepted that women doctors *would* be needed in the Australian Defence Force. At least 26 women doctors (including five majors and 21 captains) served in the Australian Army Medical Corps during World War II. Most were specialists in Australian General Hospitals, some performed general duties and administration. The first to be appointed was Lady Winifred Iris Evelyn MacKenzie.[13]

---

[10] Analysis entitled: 'Mobilisation of Medical Profession [1941]', AWM Files, War of 1939–45 Accn No. 481/2/26 Indexed 495/1 AWM 54, AWM, Canberra.

[11] The number of women (500) compared to the total number of practitioners (6,500) is notable.

[12] Minutes of the Proceedings of Conference of Deputy Directors of Medical Services, held on 20 April 1942, reported in *Paulatim*, p. 383. AWM 54/253/4/16. See also Walker, Allan, S., Chapter 20 in *Australia in the War of 1939–1945. Series 5 – Medical – Volume II – Middle East and Far East*, AWM, 1962 (reprint), p. 422.

[13] Taken from the Tait Collection compiled by Major John Thomson Tait from Department of Defence Central Registry files in Melbourne in early 1920. Supplied courtesy of the AWM.

Lady (Dr) Winifred MacKenzie (née Winifred Iris Evelyn Smith)

Once the need for women doctors was acknowledged in 1939, Lady Winifred MacKenzie became the first woman to receive a commission in the Australian Army Medical Corps.[14] Like so many pioneers, her circumstances were certainly unusual, if not unique.

Winifred Iris Evelyn Smith was the daughter of Arthur N. Smith, a journalist. She was born on 10 April 1900 and attended high school in Melbourne. She obtained her medical degree (Bachelor of Medicine and Bachelor of Surgery) from the University of Melbourne in 1924. She was not a gifted student and struggled through her medical course, initially failing seven out of 17 subjects. This required her to spend her Christmas holidays studying for supplementary examinations.[15] She was known throughout medical school as 'Win Smith'.

As a young graduate in 1925, Win went to work for Colin MacKenzie, a Victorian orthopaedic surgeon and comparative anatomist, in his private laboratories on St Kilda Road, Melbourne. Colin MacKenzie is best known for his work on the Comparative Anatomy of Australian Fauna, although during World War I he spent three years at the College of Surgeons of England cataloguing specimens of war wounds for the Army.[16] He wrote a paper on Military Orthopaedic Hospitals, that was published in the *British Medical Journal* in 1917.[17] He also served with the surgical staff of the Military Orthopaedic Hospital, Shepherds Bush, in London. On

---

[14] 'Women Officers – Australian Army Medical Corps,' 1944, AWM Files 422/7/8. War of 1939–45 AWN No. 481/2/23; Indexed 88A/2. The memorandum deals with Women Officers of the AAMC and therefore is not concerned with The Australian Army Nursing Service or The Australian Army Women's Service. See also Pearn J. H. Major Lady MacKenzie AAMC, 'The first women of the Royal Australian Army Medical Corps,' *ADF Health* 2003, Vol. 4, pp. 93–94.

[15] University of Melbourne Archives. Smith, Winifred Iris Evelyn [see also under Mackenzie [sic] Enrolment No. 190232. Student Record Card. [Giving details of student examinations in the Faculty of Medicine 1919–1924]:1–3.

[16] MacKenzie, Colin. *The Action of Muscles: Including Muscle Rest and Muscle Re-education*, London, 1918.

[17] MacKenzie W. C., 'Military Orthopaedic Hospitals', *British Medical Journal*, 1917, pp. 669–678.

his return to Australia he became an influential orthopaedic surgeon in Melbourne.

Win and Colin MacKenzie were married in 1928. Twenty-three years his junior, she gave up her medical practice specialising in orthopaedics and anaesthetics to become his research assistant and help him administer the Institute of Anatomy at Canberra. The following year Colin MacKenzie was knighted.

In 1936 Sir Colin was offered £100,000 for his unique collection of Australian flora and fauna by the Rockefeller Institution of the United States. He refused the offer and donated it to the Commonwealth,[18] along with 80 acres of bushland that is known today as the Healesville Sanctuary.[19] It was there, in 1943, that the platypus was first bred in captivity.[20]

Ten years into their marriage, Sir Colin MacKenzie died unexpectedly at the age of 62 from a cerebral haemorrhage, leaving Lady MacKenzie as a young widow with no children.

The following year, at the outbreak of World War II, the 39-year-old widow was listed as one of the 323 female medical practitioners registered in Australia. Keen to contribute to the war effort and armed with considerable administrative skill developed during her years working for her husband, Lady MacKenzie undertook full-time voluntary civilian service in the role of AAMC Staff Officer at Army Headquarters in Melbourne. She remained a volunteer for six months before she was offered a commission with the rank of Captain on 25 September 1940.

The announcement of her appointment was reported in all major newspapers around the country and caused great excitement. The press universally emphasised the "unusual circumstances" of her position – she was a widowed medical practitioner with specialist experience, proven skills in administration and who "for some months" had been working in

---

[18] 'Death of a Great Anatomist', *Sydney Morning Herald*, 30 June 1938, p. 12.

[19] Monica MacCallum, 'MacKenzie, Sir William Colin (1877–1938)', *Australian Dictionary of Biography*, National Centre of Biography, Australian National University, See: http://adb.anu.edu.au/biography/mackenzie-sir-william-colin-7392/text12831 Accessed 13 June 2013.

[20] See: 'The David Fleay Story' at http://World Warw.nprsr.qld.gov.au/parks/david-fleay/culture.html Accessed 13 June 2013.

Major Lady Winifred MacKenzie, Deputy Assistant Director General of Medical Services, A Branch, Allied Land Headquarters, in her office at Victoria Barracks, Melbourne in November 1944. *Image courtesy of the Australian War Memorial, 030214/12.*

an honorary capacity for the military. Many reports used her specialisation as a sobriquet:

> The Minister for the Army (Senator McBride) said today that appointment of women to the AAMC would be made from those who were specialists in such branches as X-ray work, pathology, ophthalmology and anaesthetics and not immediately from general practitioners. More than 1,000 male doctors had offered their services for overseas but not more than 388 had so far been appointed. There was therefore no occasion yet to utilise the services of women doctors, except in specialist capacities.[21]

Senator McBride added:

---

[21] 'Woman AAMC Officer', *West Australian*, Saturday 19 October 1940, p. 8 and 'First Women Officer in Army. Rank of Captain for Lady MacKenzie', the *Advertiser* (Adelaide), Saturday 19 October 1940, p. 14.

World War II: a corps in transition

> It would be unthinkable to denude the civil population entirely of the services of specialists, especially in big public hospitals. It was for this reason that the Army proposed to employ women specialists when necessary.[22]

Clearly, although the doors of opportunity had opened, women were still expected to take 'back seat' roles. The news of MacKenzie's appointment also made headlines in women's columns, where there was discussion not just about her pioneering role, but also her uniform:

> Of medium height and firm build, Lady MacKenzie's chief charm lies in her kindly eyes and understanding smile. From recent photographs it may be seen that she is a trim figure in her uniform which adapts the regulation pattern for military officers to feminine requirements and resembles that worn by women of the Royal Army Medical Service. The tunic and skirt are of khaki cloth, the tunic following the AIF pattern with large patch pockets, badges of rank on the shoulder tabs and unit badges on each collar lapel. Shirt, collar and tie are regulation khaki, stockings opaque and shoes tan. The officer's cap is replaced by a brimmed hat of khaki which also carries the unit badge in front.[23]

Captain (Lady) MacKenzie's roles were principally administrative within the office of the Assistant Director-General of Medical Services (ADGMS). She worked in A Branch, Allied Land Headquarters, at Victoria Barracks in central Melbourne.[24] Her duties included maintaining accurate records of the professional qualifications of the (male) doctors of the RAAMC and keeping details of their duty allotments. She also dealt with the appointment of the masseuses (physiotherapists) and provided advice on issues of women's health. On the 9th September 1941 MacKenzie became the first female medical officer to be promoted to the rank of temporary, and subsequently substantive (1 September 1942) Major, with

---

[22] ibid.

[23] 'Woman of the Month. Captain Lady MacKenzie', the *West Australian*, Thursday 9 January 1941, p. 4.

[24] Pearn, J.

the Australian Army Medical Corps.[25]

Later she received a further temporary promotion to Lieutenant Colonel and served a period acting as the Assistant Director-General Medical Services (ADGMS). At the conclusion of World War II Lady MacKenzie was one of the few women doctors to remain active in the Army Reserve, reverting to the rank of Major. She was discharged on 31 January 1947 after seven years of service. After the war her administrative medical and military skills were put to good use with the Royal District Nursing Society, where she served in leading roles, including Vice-President, for many years.

The appointment of MacKenzie stands as a landmark in the history of women in the Royal Australian Army Medical Corps. In the official history of medical services in World War II, Allan Walker records:

> The appointment of Lady Mackenzie to the Headquarters Staff was a significant step as she was the first women doctor to be commissioned. This innovation was followed by the commissioning of more women in the medical services … for whole-time or part-time duties connected with recruiting, especially of nurses, and later of members of other women's services. Later numbers of women were employed in special departments of medical units and for routine duties in base hospitals.[26]

Major Mary Thornton (née Kent-Hughes)

While Captain MacKenzie was making history as a staff officer in Melbourne, other women had travelled across the globe to contribute to the war effort. In a move reminiscent of her World War I forbears another Victorian, Dr Mary Thornton, travelled to London after she attempted to enlist with the Australian Army, only to be rejected. By the time MacKenzie received her historic commission, Mary Thornton had farewelled her husband, entrusted her 13-year-old son to boarding school, and embarked

---

[25] ibid
[26] Walker, A., p. 42.

on service with an ally that *would* accept her and her specialist skills – the British Army. On hearing the news of MacKenzie's appointment, she was already serving with British on the Nile and lamented:

> For the thousandth time did I curse the Australian military authorities whose refusal to accept service from a woman doctor had forced me across the world.[27]

Mary Thornton was schooled at Ormiston Girls School (now Camberwell Girls' School) and worked as a Voluntary Aid Detachment (VAD) member in Melbourne during World War I. In 1926, she graduated in Medicine from the University of Melbourne.

In her final year at medical school, Miss Mary Thornton married and had a son, before being granted a divorce in 1935 on the grounds of desertion. Two years later she married Dr Wilfred Kent-Hughes, an esteemed Victorian orthopaedic surgeon. He was more than 30 years her senior and had 6 children of his own. By 1939, and the outbreak of World War II, Dr Thornton had risen to become an assistant radiologist at Melbourne's Austin Hospital. She became the first woman outside England to be awarded a Fellowship of the Faculty of Radiology of the Royal College of Surgeons (FFR) – the highest qualification of its kind for radiology specialists in the British Empire.

At the time the Australian Army had no radiologists on staff and needed advice. It is difficult to appreciate today, but at the outbreak of World War II, X-rays were by no means universal, and the Army needed advice on the quantity and type of machines to use. Routine chest X-rays were not part of the standard examination for enlisting soldiers, for example, and machines were not considered to be vital equipment for deployment.[28] This was despite the use of X-rays during World War I. Although radiology had been in its infancy as a specialty, it had proven its value in revealing fractures, detecting foreign bodies and guiding surgeons in their search for pieces of shrapnel or fragments of bone.

---

[27] Kent-Hughes M., *Matilda waltzes the Tommies*, Melbourne, Oxford University Press, 1943, p. 41.
[28] Walker A.S., 'Clinical Problems of War'. *Australia in the War of 1939–45*, Vol. 1, Chapter 59, AWM, p. 668.

Having unsuccessfully applied to join the Australian Army, Mary decided to head for England, where the British War Office indicated it *was* willing to put her specialist skills to good use. She arrived in England in February 1940 to find a county already gripped by war. Hitler's forces had already conquered most of Europe, and the British Army was fully mobilised. After initial interviews at the War Office, she received her appointment as a Lieutenant in the Royal Army Medical Corps (RAMC) with the following words of caution:

> I don't know how you will get on, I'm sure. Women in the Corps – the men won't like it.[29]

The now Lieutenant Thornton was dispatched to training and then posted as a Captain to an army hospital in Surrey.

The winter of 1940 was a nightmare for Britain as German bombs rained down on its cities, particularly in the south of England. Mussolini joined forces with Hitler and England was dependent for its very survival on her Allies, waiting desperately for America to enter the war. Blackouts were routine and instruments and bandages were kept at the ready. One night a bomb landed near the hospital where Mary was working, killing five people.

Here, Captain Thornton's first military casualties came from Dunkirk, arriving bootless, with ragged clothing and suffering the effects of prolonged immersion, compounding their injuries. During the days between 27 May and the early hours of 4 June 1940, over 300,000 Allied soldiers were rescued from Dunkirk harbor and the surrounding beaches. Harnessing the full resources of Britain's navy and a flotilla of civilian vessels and fishing boats, the Dunkirk evacuation was an extraordinary feat of courage. For the novice Captain, if it were needed, it was a salient reminder of just what she had embarked upon.

Despite her concerns about being a 'colonial outsider', Mary's work was clearly accepted by the Corps. She was promoted and sent to a hospital in London. There, she found the capital in the midst of the Blitz – the familiar

---

[29] Kent-Hughes M.

World War II: a corps in transition

wail of air raid sirens meant running for shelter and being accompanied everywhere by a respirator and a tin hat.

At night there was not a light to be seen in the bomb-shattered streets. Blackout shades were kept drawn against the nightly air attacks and people fumbled between their homes and work in permanent gloom, listening always for the air-raid siren or hurried instructions from a warden. Bombs whistled their way down to earth, leaving pavements littered with broken glass, and the remains of buildings as gaping landscapes where houses, now obliterated, had stood the day before. Fires crackled away, fuelled by open gas lines, turning the darkened skies a depressing hue.

Mary's husband visited her briefly during this time in London. During World War I he had served as a surgeon with the RAMC and was keen to re-enlist. But now, aged in his 70s, his services were politely declined, leaving Mary "to carry on the job for which he was considered too old".[30]

Just before Christmas that year she was summoned by her Colonel to a depot in the English Midlands to await embarkation on active service – to a destination as yet unknown. Her husband had returned to Australia and Mary spent a lonely and melancholy Christmas; still, she was cheered by the voice of her Prime Minister, Robert Menzies, on the radio, wishing a Merry Christmas to all Australians serving abroad: "Heavens. I was glad to hear [him]," she said.

Within a few weeks she was aboard a troop transport ship, headed to the Middle East. She was fortunate when, feeling isolated as the only woman on board, her transport married up with the 2nd Australian Imperial Force (AIF). Throughout her war service she relished the chance to mix with fellow Australians. But now, heading out to Suez, pleasure took on a more poignant note:

> The joy of being welcomed by one of my husband's old students, and by nursing sisters whom I had known in Melbourne! The finding of a cousin, and the fiancé of one of my best friends! It was veritably the next best thing to going home. Literally the happiest moment of my life, as all the loneliness, depression and

---
[30] Kent-Hughes, p. 23.

homesickness of the past months gave way to the warmth and companionship of 'my [own] folk'.[31]

Little did she know that many on board, not knowing their destination either, would later be written into Australian military history as the 'Rats of Tobruk', or become prisoners on Crete, in Greece or Syria. Many of them would never return home.

For Mary, journey's end was the port of Suez with its high apricot-coloured cliffs rising from the sea. She was instantly met by the sights and smells that had astounded many Australians in khaki; mud houses, darkly veiled women and the cacophony of noises that characterise the Middle East.

Having grown up on a sheep station at Warrandyte near Melbourne, Mary was profoundly struck by the contrasts between her own beloved Australian bush and the rich history of the Middle East. After her war service she published two books, *Matilda Waltzes with the Tommies,* which recounts her own war experiences, and the fictional *The Dust of Ninevah.* In both, her astute observations show clearly how she embraced the rich mix of contemporary history and antiquity.

The arrival of a woman army doctor in Egypt was unexpected and no provisions had been made for accommodation of a female major. Seizing the freedom of her situation, Mary managed to undertake some unaccompanied sightseeing – including the Sphinx and the great pyramids – treading the very ground that had featured on so many of the postcards sent home by Australian soldiers during World War I. As an 'unruly Australian' in a British unit, she may have had more latitude as a female officer – and this included her uniform. As no-one had sent through any official orders regarding headdress for female officers, Mary took the opportunity to sport the slouch hat presented to her by a fellow Australian on the troopship. She also had her tin helmet emblazoned with a hand-drawn kangaroo and her own personal motto – 'Keep your tail up.'

Posted to 61 British Hospital, Major Thornton was initially based at Nazareth. Here she was able to travel, take leave to swim in the surf off Jaffa and to socialise with fellow Australians:

---

[31] ibid.

World War II: a corps in transition

Painting of Major Mary Thornton's tin hat emblazoned with 'Keep your tail up'. *Sourced from her autobiography: Kent–Hughes, Mary,* Matilda Waltzes with the Tommies, *Oxford University Press, 1943.*

> Here a friend of my husband's [is] commanding an Australian General Hospital just established near the orange groves, and with him four fellow students of mine and some nursing Sisters … Home news hot off the press! The Aussies never understood how I nearly fell on their necks with delight at these chance encounters; they all worked together as an Australian unit, I was a poor lone wolf … the British were consistently kind to me … but in our Corps, officers were constantly being changed from one unit to another and with an Englishman the transition from period of courteous acquaintance to 'dinkum cobber' is a long one.[32]

It was in July 1942 that she felt the separation from her own countrymen most acutely. In June, Tobruk fell. The Germans crossed the border into Egypt and casualties began to arrive from the western desert, en route to convalescent camps in Egypt. As the convoys pulled up, Major Thornton was confronted by familiar faces. On one occasion, an old friend was

---

[32] ibid, p. 101.

among the wounded – Lieutenant Colonel Raymond Anderson DSO, Commanding Officer of the 2/32nd Battalion AIF, had his arm amputated at Tobruk and was evacuated to an Australian General Hospital. Mary had been his family physician before the war, and had brought two of his children into the world. She made a hasty one-day journey to see him and was relieved to find him in good spirits. Only later, when 61 British Hospital moved north, closer to Basra in Iraq, did she receive a brief note from the Lieutenant Colonel saying how well he was feeling and thanking her for her efforts. In the same postal delivery there was a note from the Australian Matron at El Kantara who had supervised his care. It advised that Lieutenant Colonel Raymond Anderson DSO had died from his wounds.

Near Basra, the British hospital re-established itself as a tented facility in the desert. For Mary, whose job was a radiologist, this brought particular challenges. Military radiographers were required to operate their own X-ray equipment, including the darkrooms and generators required in the field. In addition, they had to find solutions to equipment breakdowns and processing issues – from breakages to poor image quality, affected by heat or dust.

Dust was everywhere. Sand floors were watered daily, but sand particles penetrated everything, finding their way into eyes, noses, mouths and machinery. The sand was an interminable grit that made its way over, under and through tent flaps, leaving a coating of khaki dust.

The heat was also excruciating. Developer tanks corroded and cooling units broke down. Clean water was always difficult to obtain and often contaminated with organic material and chlorine, leading to oxidation of the solutions used to develop X-ray images.[33] There was enormous pressure to make the first exposure a good one. Film was always in short supply and delicate X-ray tubes often broke during shipping. The X-ray machine needed ice to remain functioning and keep the developer solution cool. Otherwise the emulsion and pictures would simply slide off the film. With the aid of a neighboring Scottish engineer regiment, Mary developed a system of keeping the equipment cool and reducing the ice requirements

---

[33] Lauer, O. G., *Radiography in the United States Army during World War II*, 'Part II. Radiologic Technology', 1987, 58(3): pp. 215–223.

World War II: a corps in transition

– an ingenious design similar to the hessian-sided Koolgardie safe.[34]

Resourcefulness was a necessity, and often involved scrounging though the desert at night for precious pieces of wire, even tungsten points that could be used to repair the precious machine. Ingenuity was also a prerequisite, especially in contriving a dark room in a tent.

Major Thornton's unit possessed the only portable X-ray machine in the desert – which resulted in many trips around the neighboring Indian hospitals to take bedside images. To 'earth' the machine they:

> … passed a wire out a window to a piece of piping which was driven deep into the sand. Then an Indian poured water down the pipe just before the exposure was made. [35]

Living and working in the desert was an isolating experience, especially for a woman separated from others by nationality as well as gender. Men were brought in mangled and war-weary – their youth stripped off them by the effects of war. The omnipresent heat affected patients and staff alike – sapping energy and fraying tempers. Busy days and nights were intermingled with periods of comparative boredom and relative idleness.

As the effects of her service took their toll, so did the effects of accumulated radiation exposure. Although attempts at radiation shielding had been introduced into civilian practice in World War II, there were limited options in the field to protect radiologists or technicians from the scatter of X-rays. Radiation protection was at best inadequate – often only with sandbags.

Burning pains affected three of her fingers – no doubt the effect of radiation exposure. She was allowed 14 weeks compassionate leave back to Australia, a journey home that was epic in itself. She travelled by ship, flying boat and train through Burma, Singapore and finally to the foot of her beloved Dandenong Ranges – a place of familiar bushland and kookaburras. Her leave to return home was recorded in the *Sydney Morning Herald*.[36]

---

[34] Kent-Hughes, p. 149.

[35] ibid, p. 151.

[36] *Sydney Morning Herald*, Friday 26 September 1941, p. 3.

But her respite home as wife and mother, among the familiar landscape and scent of eucalypts, was short lived. While on leave her husband died unexpectedly, aged 76.

By the time her leave expired in 1942, the Australian Army had changed its policy to accept woman doctors. Mary could have stayed at home to mourn her husband. She could have tried once again to enlist in the uniform of her own country, but instead, recalling the words of Lieutenant Colonel Anderson in his hospital bed, "It doesn't matter what happens to you or me, Mary, it is what is best for Australia that we have to think of", she decided to return to her unit in the desert.

Saying farewell to her son in Sydney, Mary departed by flying boat from Darwin. By this stage Japan was on the offensive, and to avoid the enemy Mary's return journey was through Batavia in Indonesia. Within weeks Java was overrun, but Mary was fortunate; she made it through. Her unit was surprised that she had chosen to return, but grateful, particularly as no-one had been able to use the X-ray equipment in her absence.

The war continued on, but Mary was soon repatriated home; she was medically unwell and suffering again from the effects of her tour of duty and constant radiation exposure. Affected by a "constant weariness – when moving a finger let alone an arm or leg calls for all the effort one can muster" – she returned to Australia on board a hospital ship, her tour of duty over. Clearly relieved to be heading home, she stood astern and commented to a fellow Australian soldier that she "could almost smell the gumleaves".

After the war, Dr Mary Thornton became a visiting specialist in radiation at the Repatriation General Hospital in Heidelberg, Victoria. She was not alone as a female Australian medical specialist who had served with British forces overseas during World War II.[37] Major Joyce Wharton served with the RAMC from 1943 to 1949 as a specialist gynaecologist; her work included post-war duties in Singapore. Captain Faith Phair was appointed to the War Office, London in 1940 and subsequently undertook service in the Middle East.

When the AAMC did open its doors to women, they made significant

---

[37] Williams, Lesley, *No Better Profession: Medical Women in Queensland*, p. 160.

contributions to science as well as professional care of Australian casualties. In 1941, another woman doctor with a young son attempted to enlist with the Australian Army. On this occasion the outcome was positive. She went on to make a significant contribution to Australia's war effort in the field of medical science. Her name was Josephine Mabel Mackerras.

Major Josephine (Mabel) Mackerras

Josephine Mabel Mackerras was born at Deception Bay in Queensland on 7th August 1896 into a family of scientists. She was the daughter of Dr Thomas Lane Bancroft, a medical naturalist, and the granddaughter of the internationally recognised father of Australian natural science, Dr Joseph Bancroft. Josephine was initially educated at home and assisted her father in his research projects, which no doubt informed her knowledge of plants, animals and insects. She was allowed to spend her days exploring the surrounding bushland and the large open bays replete with mangroves, mosquitoes and sandflies. As Brisbane did not yet have a medical school, she studied at the University of Sydney, graduating in 1924.[38]

Here, she met her husband, Ian Murray Mackerras, a fellow medical graduate, and they were married in 1924 'under a large river gum on the banks of the Burnett'.[39] Ian Mackerras was a native New Zealander and had served during World War I as a laboratory technician with the Imperial Forces. He also served at Villers-Bretonneux in France where he had been temporarily blinded by mustard gas and evacuated to a military hospital in England.[40]

Jo and Ian were clearly ideally suited; they both shared a love of sailing and water as well as entomology. Jo was awarded a Walter and Eliza Hall Fellowship in economic botany and initially began her research into tick resistance in cattle and fatal diseases in freshwater fish.[41] Both Mackerras'

---

[38] Ford, E., 'Mabel Josephine Mackerras', *Medical Journal of Australia*, 1972, no. 1, pp. 604–6.
[39] Mackerras, I. M. and Marks, E. N., 'The Bancrofts: A century of scientific endeavour', *Proceedings of the Royal Society of Queensland*, 973:84, pp. 1–34.
[40] Norris, K. R., *Historical Records of Australian Science*, vol. 5, no. 2, 1981. Available at http://World Warw.science.org.au/fellows/memoirs/mackerras.html Accessed 3 October 2011.
[41] See relevant papers published by Mackerras, Mabel Josephine, cited at http://World Warw.

served on the Great Barrier Reef Committee and helped establish the Marine Research Station on Heron Island off the coast of Queensland. Unlike other couples however, they carried fish home and settled down at their microscopes, searching the smears for parasites before eating their catch.[42] Both also held pilot's licences and Jo used the Canberra Aero Club's Piper Cub aircraft to conduct field trips to collect specimens.

In 1926 their only child, David, was born and Jo retired, "with increasing dissatisfaction", into domesticity. Domestic life clearly did not suit her and it was not long before she returned to working with her husband. Theirs would become one of the most productive marital partnerships in the history of Australian science. Over their careers they worked together on a number of research projects and published 23 joint papers.[43]

A newspaper article published in 1947, after she had completed her military service, entitled *'How to Keep a Husband and a Job',* [44] provides remarkable insight into her ability to balance family and work; it established her position as a role model to women at a time when this was far from the norm:

> She said the secret was to have a housekeeper at home. "I've been very lucky," she said. "I've had domestic help nearly all my married life. I don't know how I would have managed without."

With the outbreak of war, Ian re-enlisted on 13 October 1939 and was appointed a Major in the RAAMC. He was posted to the Middle East where he served with the 2/1st Australian General Hospital at Gaza Ridge as a pathologist, also providing services in sanitary entomology. Ian was subsequently sent to North Africa to investigate an outbreak of diarrhoea. He was twice Mentioned-in-Despatches for his efforts during World

---

womenaustralia.info/biogs/AWE3718b.htm, Accessed 13 June 2013.

[42] William Lesley. Mackerras, Mabel Josephine (Jo) (1896–1971), *Australian Dictionary of Biography*, National Centre of Biography, Australian National University, Available at: http://adb.anu.edu.au/biography/mackerras-mable-jospehine-jpo-11411/text19545 Accessed 1 August 2011.

[43] Doherty, R. L. (1978), The Bancroft tradition in infectious disease research in Queensland, *Medical Journal of Australia* 1978, 2: 560–3, p. 591–4

[44] How to Keep Husband and Job. *The Mail* (Adelaide) Saturday 2 August 1947, p 2.

War II, before returning to Australia to undertake further research.

Jo was also keen to serve and enlisted as a doctor and medical scientist in November 1941 in the 2nd Military District, covering New South Wales. She was commissioned as a substantive Captain on 7 February 1942, one week before the fall of Singapore, and posted to 103 Australian General Hospital. After several transfers she was posted to Cairns in 1944 to work in the Land Headquarters Medical Research Unit, where her husband, now returned from the Middle East, was also posted.[45] The unit was led by Brigadier (later Sir) Neil Hamilton Fairley, and its principal focus was to evaluate the effectiveness of available anti-malarial drugs aimed at suppression or treatment of the disease. Urgent research on vector-borne diseases (transmitted by blood-sucking insects) was necessary and Ian Mackerras became responsible for the entomological program. He made many field trips during the war to Port Moresby and Milne Bay in New Guinea, areas where Australian troops were stationed and subject to disturbingly high rates of malaria and dengue infection.

Malaria: a common enemy

The entry of Japan into the war on 8 December 1941 focused Australia's attention on the Pacific. As Japanese forces threatened to advance from Asia into the Pacific, Australian forces were sent to Port Moresby. Unlike the campaigns of World War I in Macedonia and Palestine, this was a jungle war, but there was a common enemy – malaria.

Malaria has been a scourge of military campaigns throughout history and the Pacific War was no exception. The severe effects of the disease were first encountered by Australian forces in World War I. By the end of the Palestine campaign almost half of the 40,000 strong Desert Mounted Corps had been evacuated with illness, with 6,347 testing positive for malaria.[46] Despite this past experience, the potential for malaria to decimate Australian troops in the Pacific during World War II was underestimated.

---

[45] Ford E. Obituary. Mabel Josephine Mackerras. *Medical Journal of Australia* 1972;1:604–6

[46] Shanks, G. D., 'Simultaneous epidemics of influenza and malaria in the Australian Army in Palestine in 1918'. *Medical Journal of Australia*, 2009, no. 191, pp. 694–675.

Use of insecticide sprays and mosquito netting reduced malaria rates in camps, but these measures were not helpful for those out on the jungle tracks. Irregular supply of the suppressive drug quinine, along with the development of resistance, contributed to the increasing infection rate.

During the opening stages of the New Guinea campaign, both at the Australian base in Port Moresby and out in the jungle, malaria caused four times as many casualties among the Allies as Japanese weapons. In the first six months of 1942 alone, 1,184 cases of malaria were reported amongst 6,500 Australians. At first these were mostly due to *Plasmodium vivax* – a less virulent species of malaria parasite – but by June 1942 most were infected by the more serious and deadly *falciparum* parasite.[47] By December that year, infection rates had soared to 82 men per 1,000 troops, per week.[48]

Large numbers of soldiers were incapacitated, requiring repatriation to Australia and often prolonged recovery periods lasting up to 10 months. There were fears that infected troops returning to Australia could spread the disease across the mainland.[49]

Carried by the *Anopheles* mosquito, which bites from dusk until dawn, malaria is insidious, with fevers occurring at intervals of a few days in its early stages. As the disease progresses, crippling pain and drenching sweats would require withdrawal from the line, increasing the workload on others. The spleen, normally the size of a fist, would swell; the skin turned a sickly yellow and acquired a hue of green; victims suffered uncontrollable itching and crippling muscle and bone aches. Its worst form, cerebral malaria (caused by the *falciparum* parasite), could strike without warning, leaving its victim unconscious or to die rapidly.

The impact of malaria on fighting capacity, in an already highly contested conflict, was a source of great anxiety and the Australian Army devoted significant research funds to find ways to prevent and treat the

---

[47] Fenner, F., 'Malaria control in Papua New Guinea in the Second World War: from disaster to successful prophylaxis and the dawn of DDT', *Parassitologia*, 1998, no. 40, pp. 53–63.

[48] ibid.

[49] Condon-Rall, M. E., 'Allied Cooperation in Malaria Prevention and Control: The World War II Southwest Pacific Experience', *Journal of the History of Medicine and Allied Sciences*, 1991, 46 (4) pp. 493–519.

disease. This led to the establishment of the Land Headquarters Medical Research Unit in June 1943, under the leadership of Colonel (later Brigadier) Fairley.

Here, the first research experiments undertaken by Jo Mackerras involved collecting pupae of the *Anopheles punctulatus* mosquito brought from New Guinea to Cairns where they grew to maturity in purpose-built laboratory facilities.[50] At the peak of operations, 20,000 larvae per week were transported to Jo's lab.

Cairns was an ideal environment. It was close enough to airfields to transport live mosquito pupae; it was also the principal recreation and retraining area for troops temporarily withdrawn from New Guinea. Here, tanned Australian servicemen fresh from fighting in PNG became experimental guinea pigs.

Servicemen carrying *P. vivax* or *P. falciparum* in their blood were exposed to the adult *punctulatus* mosquitoes. These mosquitos were then used to infect malaria-free volunteers from across Australia (research that would be unlikely to pass any ethics committee approvals today).[51] This allowed the team to test the effectiveness of a new antimalarial drug, *Atebrin*, which had been developed by German chemists just before the war.

Over the next few years Jo's meticulous research involved more than 1,000 human volunteers in extensive experiments at Cairns and Rocky Creek, on the Atherton Tablelands. The research into malaria control and the use of suppressive medication was of urgent strategic importance to Allied forces in the Pacific.[52] With these experiments the team not only demonstrated the effectiveness of *Atebrin* in preventing or suppressing malaria, but infected subjects could effectively be 'cured' and subjected to the demands of jungle warfare without recurrence of their symptoms.[53]

---

[50] Fenner, R.

[51] The Nuremburg code for human experimentation, to which Australia is a signatory, requires explicit voluntary consent for patients, but was not drafted until Aug 9, 1947. It is doubtful if such a code, even if known, would have factored highly, given the extent of the losses at the time.

[52] Beadle, C. and Hoffman, S. L., 'History of Malaria in the United States Naval Forces at War: World War I Through the Vietnam Conflict', *Clinical Infectious Diseases*, 1993; vol. 16 (2), pp. 320–9.

[53] Sweeney, A. W., 'The malaria frontline. Pioneering malaria research by the Australian Army

In order to test the drug's effectiveness in fighting environment, the men were put under stress – chopping wood for days at a time in the Queensland tropical heat, swimming upstream until they sank with fatigue, and being forced to march through the Tablelands. Aircraft environments were simulated by placing volunteers in refrigerated rooms in full flying suits or into depressurisation chambers for two hours per day. The results were conclusive – not one developed malaria.[54] Both the *vivax* and *falciparum* forms of malaria were suppressed and *Atebrin* became the Australian Army's drug of choice.

It was in her Cairns laboratories, driven by failing supplies of larvae from New Guinea, that Jo Mackerras became the first person in the world to establish a breeding colony of *Anopheles punctulatus*. This was a significant breakthrough that reduced scientists' reliance on mosquito larvae being brought from overseas.[55] This single advance opened up opportunities for a whole new range of experimental work, in particular the study of resistance and prevention measures that occupied Jo for the rest of the war.[56]

In her three years of active military service, she handled: "233,000 engorged mosquitoes, undertook 38,000 dissections of malarial mosquitoes and supervised more than 20,000 infectious bites over 1,000 human volunteers, some of them several times."[57]

Her most trying experience came when her insectary – housing her 'brood' – was threatened by a hurricane: she was on standby for 24 hours armed with insecticide, ready to destroy her mosquitoes in case they were released into the community. She was acutely aware of the consequences if infected mosquitoes were to escape; she was also aware of the repercussions for her team's research efforts, should they be destroyed.

It is a testament to her work, as part of Brigadier Fairley's team, that the

---

in World War II', *Medical Journal of Australia*, 1997 March 17, 166(6), pp. 316–9.

[54] Fenner, R. and Sweeney, A.W., 'Malaria in New Guinea during the Second World War: the Land Headquarters Medical Research Unit', *Parassitologia*, 1998, no. 40, pp. 65–68.

[55] Mackerras, M. J. and Lermerle, T. H., 'Laboratory breeding of *Anopheles punctulatus*', *Bulletin of Entomological Research*, 1949 May, no. 40(1), pp. 27–41

[56] Fenner, F., 'The Impact of the Bancroft Kindred on Australian Medical Science', *The Bancroft Tradition*, Eds Pearn, J., and Powell, L., Brisbane, Amphion Press, 1991, p. 59.

[57] Ford, E.

World War II: a corps in transition

Captain Mabel Josephine Mackerras (far right), of the malarial research group at the 5th Australian Camp Hospital in Cairns, dissects mosquitos and checks for infection. *Image courtesy of the Australian War Memorial*, 057738.

rate of malaria in highly infected areas dropped from 740 cases per 1,000 men in December 1943 to 26 cases per 1,000 by November 1944, less than one year later.[58]

Professor Frank Fenner, a former medical officer with the AAMC, and one of Australia's greatest biological scientists of the time wrote in 1991:

> Her results were of great importance to the war effort, since they established a scientific basis for chemoprophylaxis that was eventually to transform into a minor problem what had threatened to be a disease that would totally disable the Australian and United States forces in the field in New Guinea … [This work] at the Land Headquarters Medical Research Unit on Malaria was not only of great importance to the war effort, but also contributed greatly to the understanding and pathogenesis of malaria.[59]

---

[58] Spencer, M., 'Malaria and the Land HQ Medical Research Unit 1943–46', MJA, 1994, 161(4), p. 78.

[59] Fenner, F. p. 59.

After her discharge from the Army in 1946, Jo continued to work with her husband at the CSIRO laboratories in Yeerongpilly near Brisbane. Her pioneering work in medical entomology created a legacy that she continued after the war – she went on to discover that cockroaches transmit *Salmonella* and were a vector for childhood gastroenteritis.[60]

Jo also had a natural ability to nurture and encourage those working with her. During the war she prepared gin slings in 1,000cc cylinders and served them to her team when the day's work was done.[61] Throughout her career, her house was always open to young scientists and researchers. She became President of the Medical Women's Society and the Women Graduates Association; she was also elected a fellow of the Royal College of Pathologists in Australasia and the Australian Society of Parasitology. A meticulous and dedicated scientist, her research resulted in over 80 papers. She is said to have possessed:

> … a serene charm, a placid smile and a shy, self effacing manner. Quietly and unobtrusively, she fostered young scientists and won the esteem of her senior colleagues.[62]

Josephine Mackerras died on 8 October 1971 and is buried in Canberra cemetery. Her death ended 46 years of married life; she worked shoulder to shoulder with her husband for approximately four decades in that time. The research efforts of Jo and her husband during the war made a major contribution to the war effort and a lasting contribution to the control of malaria.[63] Her portrait, by Nora Heysen, commissioned as Australia's first female war artist, hangs in the Australian War Memorial, Canberra.[64]

---

[60] Mackerras, I. M. and Marks, E.N., 'The Bancrofts: A century of scientific endeavour', *Proceedings of the Royal Society of Queensland*, 1973, no. 84, pp. 1–34.

[61] Ford, E.

[62] Williams, Lesley. See footnote 41.

[63] Sweeney, A. W., 'The malaria frontline. Pioneering malaria research by the Australian Army in World War II', *Medical Journal of Australia*, 1997;166(6): 316–9.

[64] Prosser, C. L., Clark, I. A., 'The war on malaria and Nora Heysen's documentation of Australian medical research through art between 1943 and 1945', *Medical Journal of Australia* 2011:194 (8):418–419.

World War II: a corps in transition

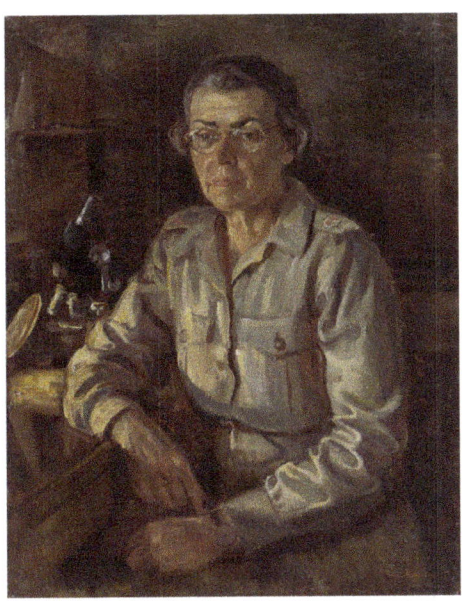

Portrait of Mabel Josephine Mackerras, senior entomologist, Land Headquarters Medical Research Unit, Australian Army Medical Corps (AAMC). Painted by Australia's first female war artist, Nora Heysen, the portrait hangs in the Australian War Memorial.
Image courtesy of the Australian War Memorial, ART24395.

SCIENTIFIC OFFICERS AND ALLIED HEALTH

Women's roles in blood transfusion services

The science of blood transfusion was still in its infancy at the outbreak of World War II. Although it was used extensively during the later stages of World War I, and played a significant role in saving lives, there were still technical problems. There was also the risk of reactions to transfusion that could be fatal, and transfusion was not widely used.

In the interwar period there were two major advances in transfusion. One was the development of blood banks for storage; the other was the establishment of a national transfusion service through the Australian Red Cross. In Australia, the first blood bank for general use was established at the Royal Melbourne Hospital, with essential research undertaken at the nearby Walter and Eliza Hall Institute.[65] When World War II broke

---

[65] Walker, 'Clinical Problems of War'. See Chapter 34: 'Blood Transfusion'.

out, there was only one source of blood serum for the entire nation: the Commonwealth Serum Laboratories (CSL) in Melbourne. Work was underway to create national stockpiles for civilian needs in the event of an air attack on Australia.

Blood contains red and white corpuscles or cells, suspended in fluid. This straw-coloured liquid, known as plasma or serum, is mainly composed of water, blood proteins and inorganic electrolytes. It transports glucose, fats, hormones, metabolic products, carbon dioxide and oxygen throughout the body; it also carries the essential ingredients for blood to clot. In times of war and conflict, blood supplies play a central role in the treatment of casualties. Blood serum has particular importance in the aftermath of explosions. It is used to replace the vast amounts of fluid and protein lost through burned skin.

In 1940, the Australian Army's first mobile blood bank met a disastrous fate when it was transported, by ship, to Australian forces in the Middle East. Captured by the Germans, the ship and its cargo were later sunk by the British off the French coast.[66] At home, there were also concerns about the risks of transportation, and the nation's reliance on blood supplies from CSL in Melbourne.

That same year, Dr Cyril Fortune, an early proponent of blood transfusion, recognised Australia's 'tyranny of distance' concerning blood supplies. He was keen to establish a second transfusion laboratory in Perth, rather than rely on lengthy transportation from Melbourne. As the war moved into the Pacific he grew even more concerned that Perth could be isolated and supplies of blood serum could become unavailable.

After Fortune had successfully lobbied for a Perth-based service, a young Western Australian woman, 26-year old Jean Kahan was to play a key role in establishing the unit.[67] Pivotal in determining the war's outcome, blood transfusion was meticulous work; it was an area of medical science where women played a major role.

---

[66] See: 'History of Blood Services 1939–1945', Australian Red Cross Blood Service. Available at: http://www.donateblood.com.au/about-us/history-blood-services/1939-1945 Accessed 16 June 2014.

[67] See: chapter 19 in Adams-Smith, Patsy, *Australian Women at War*, Penguin, 1984.

World War II: a corps in transition

Lieutenant Jean Kahan

Jean Kahan served with the Australian Army for six years during World War II (from 1940–46). Before the war broke out, Jean had graduated in science with a major in zoology from the University of Western Australia. Like many women, she had found it difficult to find a scientific post, and was using her skills teaching mathematics and biology.

As the niece of one of the original ANZAC soldiers, Jean was keen to enlist, but initially found it difficult to find her 'niche'. With her background in science she was asked to work in the mustard-gas section of munitions research centre in Maribyrnong, Victoria.

It would, however, be her relationship with the far-sighted Dr (later Major) Cyril Fortune that determined her real contribution to the war effort, and her experiences are typical of many Australian women who worked in the emerging blood services. In the years before the war, Jean had worked as a volunteer with Dr Fortune, collecting blood samples from enlisting soldiers at Perth's Swanbourne Barracks. Here she was taught the fastidious work of 'cross-matching': identifying which of the four main blood groups the men belonged to – A, B, AB or O. Small samples of blood from the donor were dropped onto a glass slide and carefully mixed with 'test solutions' from each of the other blood groups. The slides were then examined under a microscope to see if any 'clumping' or agglutination occurred, a sure sign of a potentially fatal transfusion reaction. Sometimes the blood would clump despite being from the same group and, supposedly, 'compatible'. Only later in the war was a second major factor identified – the Rhesus factor.

Despite Dr Fortune's earnest encouragement for her to stay in Perth and help set up a transfusion service there, Jean felt that her duty lay elsewhere. She went where she felt she was most needed – Maribyrnong in Victoria.

Munitions Work

Throughout World War II women worked producing munitions, explosives, weaponry, optical equipment and respirators. While most workers at the factory were men, the number of female workers in the years 1939–45 rose as high as 3,197.[68]

The consequences of exposure to poisoning when working with mustard gas were immense, so stringent blood testing was required. Jean was anaemic, which should have excluded her from such work, but after a few months of injections and medication, Dr Fortune reluctantly certified her as 'fit' and she headed east.

At Maribynong, special laboratories were set up to test equipment, such as gas masks. The clothing of servicemen was also tested – for example, seams were subjected to mustard gas to establish that they would not let the gas through.[69]

Initially, research relied entirely on the experiences of British troops exposed to chemical agents, but it was soon realised that Australians were fighting in tropical environments that were markedly different. The effects of mustard gas are much more severe in the heat. The chemicals used to impregnate clothing, and make it resistant to the gas, were also absorbed by soldiers' skin. This converted their haemoglobin into the deadly poisonous methhaemoglobin, preventing oxygen being delivered to the cells; resulting in a form of 'cellular aspyxiation'.[70]

Jean was working in the laboratory at Maribynong on 7th December 1941 when Japanese planes attacked Pearl Harbour. War suddenly came much closer to Australian shores, and the need for essential transfusion capabilities in key cities, including Perth, became more urgent.

With encouragement from her superior, now Major Fortune, Jean negotiated the difficult process of leaving Maribyrnong. Released from her duties on the understanding that her work in Perth would be equally essential to the war effort, Jean went about learning as much as she could to take essential knowledge and supplies back to Western Australia. She spent time at military instillations to observe how the staff were organised

---

[68] Kahan, Jean, *The Cream Separator and the bleeding room*, self-published, 2006, p. 2.
[69] ibid.
[70] Gillis, R. G. (ed), 'Dark Days for Australia', *The Gillis Report: Australian Field Trials with Mustard Gas 1942–45,* Peace Research Centre, ANU, 1985. Available at: http://mustardgas.org/The-Gillis-Report-(1985)-Australian-Field-Trials-With-Mustard-Gas-1942-1945.pdf, Accessed 14 June 2013.

and trained, and she worked at the Walter and Eliza Hall Institute (WEHI) to learn the procedure for making, storing, packaging and transporting serum.

Jean saw her first blood collection at WEHI – she described it as an experience where science met practicality:

> The first blood transfusion I saw, I was told to hold the bottle while they drained the blood from a man. I sat holding the bottle in my hands and the warm fluid coming from a body – I felt things going black and I said to myself that I wouldn't be much help in a war if I fainted at the first trial, so I got firm and pulled myself up. It was the just the warmth that was strange to me.[71]

Here she was paid by the Army, but had no official position. There was much to learn: the cleaning of equipment, sterilising, making up sets for collecting and transfusing blood, testing specimens grown on agar plates to ensure that the work was completely sterile at each stage. The key to the process was to separate the elements of blood to allow them to be stored and transported. Jean described the process of separation using an ordinary domestic Alfa Laval cream separator:

> To prevent clotting, an anticoagulant such as citrate or heparin is added to the blood specimen as it is obtained. The specimen is then centrifuged to separate the plasma from the blood cells. The centrifuge used was a cream separator, driven by turning a handle as farmers' wives have always done. When we were centrifuging blood, the cells were ejected at the 'cream' spout and the plasma at the 'milk' spout.[72]

Plasma contains fibrin, which is removed by adding a chemical, leaving fluid serum. The serum was then filtered and stored. Serum had to be transported in glass bottles to wherever it was needed. In Australia alone these distances were vast, but so too were the distances to Australian troops and field hospitals in the Pacific and the Middle East.

---

[71] Kahan J.

[72] Kahan, p. 3.

Specially shaped 'Soluvac' flasks were filled to capacity to allow no space that would damage the valuable contents by shaking; they were then sealed with cellophane and wax. The bottles were tightly packed in lead-lined boxes with ice. Blood components have only a short useful life and need to be treated well – even over rough terrain to get to the dressing stations in New Guinea and Egypt where it was needed.

Having learnt all she could, Jean travelled west with her newfound knowledge and with valuable supplies. Like many things, scientific equipment was in short supply in wartime. She had obtained whatever she could lay her hands on, including a complete blood transfusion set, filters, stainless-steel piping and glass tubes. All of this accompanied her back to Perth in the luggage rack on the train, carefully labelled 'EGGS – HANDLE WITH CARE'. This avoided attention, but caused some irritation to her fellow passengers.

The Perth transfusion unit was set up in the laboratory of Hollywood Military Hospital. A team of medical orderlies (known as Voluntary Aid Detachment members or VADs) were enlisted to help. The VADs were women from a range of backgrounds and for some it was their first job. Others had held medical jobs in the previous war. They played a vital role, helping blood donors, cleaning and autoclaving equipment and washing and sharpening the transfusion needles by hand.

Maintaining the equipment and finding spare parts to keep the service going was challenging during a war. Ever resourceful, Jean eventually found that the Perth-based Swan Brewery used similar equipment in their filtering process and were willing to help. A patient, who had experience in glass-blowing, was also enlisted to help with making some key glass pieces.

At this time Jean also had the opportunity to participate in some of the early studies on the newly discovered Rhesus factor. First discovered in Rhesus monkeys, the factor proved to be the major cause of previously unexplained transfusion reactions; incompatibility between Rhesus positive and negative blood could also be a problem for mothers expecting babies. Rarely seen today, it could lead to the delivery of 'blue babies' and some did not survive. At one point, her zoology training was called into play and Jean worked with a local doctor, Dr George Kelsall, to administer anaesthetic to a Rhesus monkey at the South Perth Zoo as part of research into the problem.[73]

By now the Americans had entered the war, and submariners were frequent visitors to Fremantle where the American Navy had established a base. Jean and her team worked closely with American marine medics to ensure that each man was blood grouped and that serum was available to supply any ships coming into port. Two U.S. Marine blood technicians, Angel Flores and Lyn Cooper, were assigned to Jean's unit to help. They remained with the unit for two years.

Jean still had no official Army position, so a frustrated Major Fortune threw his car keys over to her and said "Go and get yourself enlisted". The outcome was an appointment as Lieutenant with the Australian Army Medical Corps and a trip to the tailor to have a uniform made. Jean was 26 years old. Although she served with the Australian Army for the rest of the war, Lieutenant Kahan's time wearing AAMC accoutrements was limited. Enlisted on 17 October 1942, with service number WFX37673, the vagaries of the management of female officers saw her subsequently transferred to the Australian Army Medical Women's Service (AAMWS). Only female doctors retained their positions with the AAMC.

Jean in the RAAMC

By 1942, blood and serum preparation units had been established in each capital city in Australia. These were Army units, working in conjunction with the Australian Red Cross. After setting up the unit in Perth, Jean was posted to the 2nd Blood and Serum Army Unit located within the Royal Sydney Hospital. For troops serving in the Pacific, serum was transported via hospital ships to Brisbane initially, where it was loaded on to Sunderland flying boats heading to Moresby and on to Australian military hospitals in New Guinea. There is no doubt that many Australian lives were saved through the provision of serum, particularly during the Pacific jungle campaigns.[74]

Later, in Brisbane, Lieutenant Kahan took over as the officer in charge

---

[73] Kahan J.

[74] Walker, 'Clinical Problems of War'.

of the 1st Australian Blood and Serum Unit, situated in a requisitioned house on Alice Street in the city. Besides the scientific work of the unit, Jean was also responsible for administration: pay, sick leave and vehicle maintenance all fell within her purview. She was the only female officer in the unit, and found her relationship with the Deputy Director of Medical Services (DDMS) difficult at times. Jean was not only working as a scientist, but as the leader of the small team. She took her responsibilities as an officer seriously:

> One of my girls, a VA, had seen her husband for two weeks between him returning from the Middle East and his departure for New Guinea. A few weeks later I had the job of telling her he was dead. It was an officer's job. It had to be done. There was no easy way to do a thing like that.[75]

Jean continued to run the unit until the end of the war. On 15 August 1945 Victory in the Pacific (VP) Day was marked and, like every city in Australia, the streets of Brisbane were crowded with people singing and dancing, awash with relief at the end of the war and the prospect of return to normal life.

But Jean's job was not quite over. She remained in the Army to oversee the transfer of her unit to the civilian hospital and to train her replacement. The unit was re-established at Holland Park Military Hospital where, once in smooth working order, it was taken over by the Australian Red Cross. Finally, Jean made plans to return home.

On Christmas Day 1945, Jean boarded a Lancaster bomber. Seated on the floor, with her own supplies of canteen food, she was flown to Melbourne – the first stop on the long journey back to Perth and her eventual discharge from the Army. She provided four years of valuable service, and in that time she traversed the continent from east to west a number of times.

The Laval cream separator that Jean brought from Melbourne in 1942 remained in service for 35 years. It now stands in the entrance hall of the Red Cross Building on Wellington Street in Perth.

---

[75] Adams-Smith, P., p. 206.

CHAPTER FOUR

Betwixt and between: allied health care and the struggle for recognition

There is now practically no activity in which women have not been engaged ... the recent war found women in all countries doing the hard manual labour usually reserved for men, and there is not a branch of sport wherein they have not distinguished themselves. Is it any wonder I ask: 'Wither woman next'?
  Dame Constance D'Arcy, Deputy Vice-Chancellor, Sydney University, 1946[1]

The surgeon can do so much to re-form the shattered limb, but it is may be a useless encumbrance if it does not function. This is the part where the magnificent work of the physiotherapist is so important and should commence as soon as possible from the time of injury and on awakening from the anaesethetic.
  Donald Beard AM, RFD, ED, Colonel RAAMC[2]

While the status of female doctors progressed during World War II and they found legitimacy as medical professionals, a lack of recognition still plagued women in the allied health professions.

By December 1941, "all persons with specific qualifications in any

---

[1] Malloch, N. W., 'Wither Women Next?' Supplement to *The Argus*, June 12, 1946, p. 5.

[2] Wilson, Honor, C., Physiotherapists in war: the story of South Australian physiotherapists during World Wars I and II, Japan-Korea and Vietnam, Hazelwood Park, SA, 1995.

branch of science or medical knowledge ancillary to medical science"[3] were permitted promotion to the rank of Lieutenant in the Australian Army. Far from providing clarity, however, this policy sowed the seeds of discontent and discrepancy. It was also unclear where they 'belonged'. Some women were reallocated from the Australian Army Medical Corps to the Australian Army Medical Women's Service and vice versa – issues of equity became mired in gender and dispute regarding the nature of medical professions. The situation was not resolved until 1988 when all medical personnel, other than nurses (who had their own corps and historical context), came under the auspices of the Royal Australian Army Medical Corps, regardless of gender.

PHYSIOTHERAPY AND OCCUPATIONAL THERAPY

Although masseuses had served on board hospital ships during World War I, returning with the wounded from Europe and the Gallipoli peninsula, it was not until World War II that physiotherapy established itself as a professional branch within the Australian Army Medical Corps.[4]

In 1935, with the threat of conflict returning to Europe and the Pacific, the Australian Massage Association[5] approached the Army's Director-General of Medical Services (DGMS) to consider formation of a Military Massage Unit. The association's branches in New South Wales and Victoria also lobbied for the inclusion of a massage section within the AAMC.

The DGMS was prepared to take on masseurs, but the issue of *masseuses* was, not surprisingly, more problematic. His response was:

> In regard to the Massage Reserve, which at present provides
> for the appointment of masseurs only, it is intended to make
> provision for the appointment of qualified masseuses, but the

---

[3] Walker, Allan, S., 'Medical Services of the Royal Australian Navy and Royal Australian Air Force with a section on women in the Army Medical Services', Australia in the war of 1939–1945, Series 5 – Medical, Volume IV, ed. Allan S. Walker et al, Australian War Memorial, 1961: pp. 422–423

[4] ibid.

[5] The civilian physiotherapy association in Australia was known as the Australian Massage Association until 1944, when it was renamed the Australian Physiotherapy Association.

matter is held in abeyance pending the return from abroad of the Matron-in-Chief with whom the question will be discussed.[6]

After much discussion, it was decided that:

> While masseurs would be enlisted to the AAMC, masseuses would only be appointed to the Massage Service in a similar manner to nurses appointed to the Australian Army Auxiliary Nursing Service. Further to this, while they enjoyed the 'privileges and courtesies of commissioned rank', they held only the equivalent rank of Sergeant.[7]

Ironically, Major Winifred MacKenzie, the first female officer in the RAAMC, was given the task of appointing masseuses to the Massage Service, in accordance with instructions from the DGMS HQ in Melbourne. By 9th March 1941 the first masseuses had been enlisted for service with the 2/2nd AGH Southern Command attached to the 6th Division AIF.

To distinguish them from their nursing colleagues, the women's uniforms comprised a pale-blue blouse with distinctive cherry shoulder tabs and matching tie.[8] This outfit was complemented by a felt hat and shoulder capes identical to those of the nurses, except in Saxe blue, rather than red. Early in 1943 the distinctive navy blue uniforms were replaced by khaki outdoor dress, and for the tropics, long trousers and long-sleeved shirts.[9]

During their preparations, the women were issued their uniforms and kit bags; they underwent medical examinations and vaccinations, and spent a few days in camp to be introduced to Army life and discipline. Finally, in 1941, the Australian Army's first female physiotherapists were dispatched to the Middle East. They served with Australian and British General Hospitals in Eritrea, Greece, Crete, Malaya, Egypt, Gaza, Malaya and Ceylon.

---

[6] Wilson, H. p. 14.

[7] Liebich, Genevieve, 'Physiotherapy: the development of a "new practice" and the challenges of the early years', *ADF Health*, June 2008; 9(1):43–46.

[8] Each of the Australian Army Corps has colours that typify their embellishments. The corps colour of the RAAMC is 'dull cherry'.

[9] Wilson, H. p. 17.

Among the first women to deploy was Miss Honor Wilson from Adelaide. It was a colourful sight on the pier as the RMS *Strathaird* prepared to sail – naval flags, streamers and the bustle of loading last-minute provisions competed with the farewell wishes of loved ones. The women, dressed in working uniforms, lined the decks: there were physiotherapists in navy-blue capes, the nurses' veils billowed in the sea breeze as the land slowly slipped from their view.

The hospital ships were escorted in convoy to the Suez, via Fremantle, Colombo, Aden and finally Port Tewfik. It was a long journey. Life on board settled into a routine of PT classes before breakfast given by the physiotherapists followed by sports such as quoits and deck tennis; there were also life-boat drills, air raid drills and lectures given on medical matters. Fortunately, the women had been allowed to pack a white tennis frock and bathers. At night the ship was blacked out, and those on board became accustomed to the warm air, heavy with salt, and the vast expanse of ocean.

Arriving early in the morning, the women disembarked with the troops. They faced a long, hot and dusty train journey to Gaza in Palestine, where they joined the military hospitals of the 2nd AIF. Here, they were 'attached' rather than 'enlisted' – while they enjoyed all the privileges and courtesies associated with 'officer status' they, like their forebears in World War I, were not commissioned. Their pay was 8/6 per day, to be increased to 9/4 per day on embarkation.

As was the case in World War I, issues of rank and pay became a source of controversy. Male physiotherapists were commissioned as Lieutenants and earnt 16/- per day, more than double the rate paid to female physiotherapists for the same work. After considerable debate, the DGMS decided to appoint female physiotherapists at the rank of Lieutenant, with equivalent pay rates to their male counterparts. However, far from ending discontent, there was further trouble as more discrepancies emerged regarding pay rates for women serving with the AAMWS.[10] Through 1942, various societies and professional bodies weighed in on the argument. There were obvious inconsistencies between the pay and conditions of

---

[10] See appendix: Chronology of Medical Women in the Australian Army.

Allied health care and the struggle for recognition

Physiotherapist Alison McArthur-Campbell with Sister Trembath, AANS, aboard the British hospital ship *Dorsetshire* en route to Tobruk. *Image courtesy of the Australian War Memorial, donated by E. Fussell.* P01348.001.

'senior service' AAMC and 'scientifically trained' professional women, as opposed to 'untrained women' of the AAMWS.

While debate about working conditions was ongoing in Australia, in the Middle East and elsewhere the masseuses of the Australian Army simply got on with their job. The role that physiotherapists undertook in the treatment of postoperative conditions saw them become valued and accepted members of the medical team. Their work included a wide variety of tasks depending on their posted location. Those working with the AGHs had regular exposure to war injuries: they worked with plasters and in orthopaedic wards, applying splints and encouraging function and exercise. Physiotherapy treatment was often used after back, chest and plastic surgery, especially in the treatment of burns and open wounds.[11]

A typical example was the 2/1st AGH, a tented hospital surrounded by barbed wire in Palestine. During the months of the Levant, from March

---

[11] Bentley P. The path to professionalism: physiotherapy in Australia to the 1980s. Melbourne: Australian Physiotherapy Association, 2006.

to May, the days were warm and the nights were cold. Tent flaps could be rolled up to provide a breeze, but the clay floors need to be watered as the hot desert winds brought fresh layers of sand. Here, a special thoracic unit was established and physiotherapists played a vital role in assisting breathing exercises and normal function. Working days were long, often up to 14 hours per day.

Soldiers with burns to their limbs or chest would often undergo 'escharotomies' – long incisions made by the surgeons to open the wound and stop the burn scar from acting as a tourniquet and constricting the blood supply to the limb or restrict breathing. Although painful, the burned tissues needed massage and pressure to prevent them becoming hardened. Burned joints required constant movement to prevent contractures – the physiotherapist's task.

Despite the experience of World War I, no provision had been made for outfitting physiotherapy departments in World War II.[12] The Red Cross helped where it could, but improvisation was essential,[13] as Alison McArthur Campbell (referred to later in this chapter) noted in her service diary:

> 2.4.41   Polio, [JW] aged 22 Royal Horse Artilliary [sic] "We lengthened John's bed (he was 6'4 ¼ ") fixed up and [sic] arm board and put a munitions box tied on with rope at foot of bed to prevent foot drop. Muscle charted him."
>
> 3.4.41   Physios used hot sand for alternative to Infra Red [sic], diathermy etc if required."[14]

---

[12] A situation that would repeat itself again in Vietnam.

[13] Leibich, G.

[14] Extracts from the Service Diary of Miss Alison A. Mc Arthur Campbell 9.1.40–3.3.43. Held at Australian War Memorial AWM – PR90/022.

Allied health care and the struggle for recognition

Orthopaedic cases were also frequent – physiotherapists would help patients to use crutches or walking aids correctly while they were waiting for broken limbs or shrapnel wounds to heal. They would also manage their plaster casts – these had to be prepared and applied with care and often required adjustment. Soldiers with pelvic fractures could require particularly complicated plasters and traction. Insidious particles of fine desert sand would find their way under the cast and irritate if the plaster was not properly applied. At times, the cast would be removed to reveal growing ulcers where rubbing had eroded through the underlying skin.

Plaster casts occupied endless hours of the physiotherapists' time. They were applied as roll bandages and made wet to activate the gypsum, which sets hard on drying. Timing was crucial as there is only a small window of time to ensure that the setting plaster is smooth and correctly applied before it hardens. All plaster bandages were hand rolled, wrapped in single

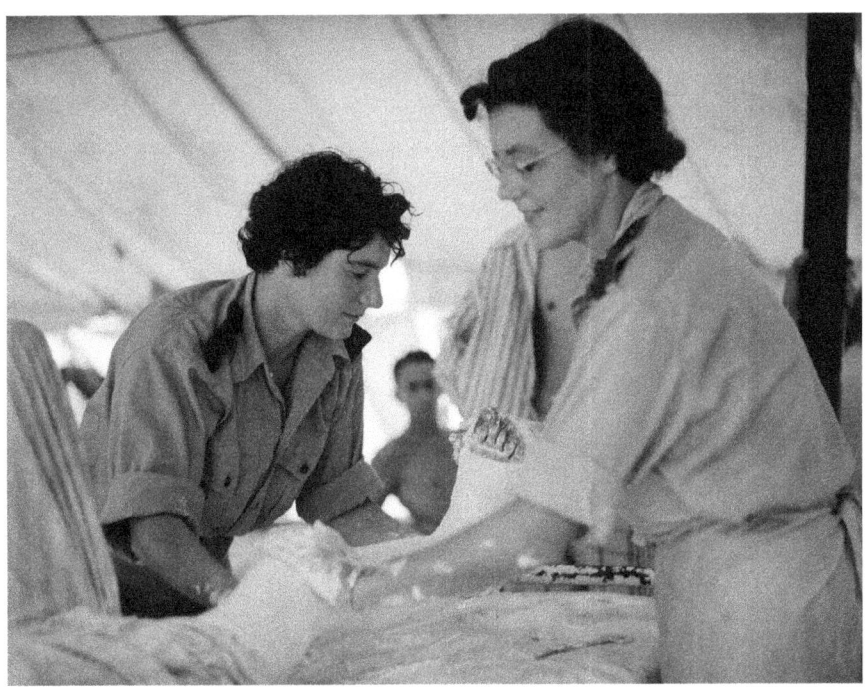

Physiotherapists Lieutenant J. Dickson and Lieutenant C. McLean work on the leg of a civilian patient at the 2/11th General Hospital at Madang, New Guinea in 1944.
Image courtesy of the Australian War Memorial, 075542.

layers, dried out in ovens and then packed in air-tight tins.[15]

Later in the war, when physiotherapists deployed during the Pacific campaign, the task of plaster maintenance was particularly challenging. It was hard to dry them out and the humidity would often soften the plasters, causing them to crack or simply become soggy and ineffective.

The physiotherapists also undertook an important function in remediating war injuries. They would apply splints and treat nerve injuries, they encouraged muscular function and showed soldiers how to care for their feet, they supervised and supported the rehabilitation for seriously injured soldiers. This work would be continued on the hospital ships back home.

Alison McArthur-Campbell's diaries record the workload and the conditions in North Africa in early 1941, shortly after the Germans had recaptured the town of Benghazi from the Allies. The following extracts begin on the day that the city fell to the German Afrika Corps, under the leadership of Lieutenant General Rommel:

4.4.41	800 patients in the hospital and coming and going all day, and a wild rush of head and facial wounds. 84 patients in ward 1 by lunchtime. Patients coming in from Derna and the all the way down the coast with appalling wounds.
6.4.41	Getting patients organised to go down to Alexandria on the hospital ship. That evening told to pack, kit bag only, and ready on call to leave.
7.4.41	Did not sleep much – noise of battle, all night trucks and tanks and cars and troops moved up and down. Awoke at 6 a.m., scrap breakfast, everything packed and in hall. Every man who called at the hospital had a fresh story concerning the size and speed of the

---
[15] Walker, A.S., p. 423.

> German army... Except for one small mine sweeper the harbor was empty. The hospital ship should have been in in the night and loading. About 10.30 a.m. Charlie the Orderly came in having spotted the ship. The relief was amazing. All round there were signs of intensive activity – ambulances and trucks coming up/down at the hospital. Tank traps were being prepared with old Ity [sic] trucks, one at our own gate. At 6 p.m. the patients having all been loaded, ambulances came and we drove to the wharf (i.e. all the women personnel from the hospital).
>
> 8.4.41 Spent morning loading in Tobruk harbour, as more and more patients came in. About 10.30 a.m. left the harbour. About 12 noon fouled a buoy rope and spent hours watching rope being freed...
>
> 10.4.41 Gaza. Palestine. It is hard to believe and terribly hard to realize that we have arrived here safe and well, while our men are fighting for their lives out there. After a night at sea we came into Haifa harbour. Said Good Bye [sic] to John (new polio). He was nearly weeping poor lad.
>
> 15.4.41 Had first real bath since 18th March.

As the war progressed and hostilities increased, some female physiotherapists came under fire and required evacuation. Immediately prior to the fall of Singapore, for example, a number of female physiotherapists and nurses were evacuated by sea to Australia. The physiotherapists boarded the *Empire Star*, which sailed on 11 February 1942. Designed as a cargo ship to carry 24 passengers, dire necessity dictated that more than 2,000 people were crammed into its holds, berths and massed across its decks. Among the evacuees were Australian, British and Indian nurses, British troops and civilian women and children. During its passage the ship was heavily bombarded by 92 Japanese bombers. Despite massive damage and huge losses of life, the *Empire Star* somehow limped to safety in Batavia. Vital repairs were carried out and the ship, incredibly, reached its destination in Australia without further incident.[16]

Although it was a harrowing journey, all of the physiotherapists on board the *Empire Star* returned home safely. Some of their ill-fated nursing colleagues were not so fortunate. Of the 65 nurses who embarked upon the SS *Vyner Brooke*, 12 were drowned when the ship was sunk in an aerial attack and 21 were murdered by Japanese troops during the Bangka Island massacre.[17] Only one nurse survived, Sister Vivian Bullwinkel, who hid for days in the jungle. Of the 32 nurses taken prisoner, only 24 survived until the end of the war.

WAR REACHES AUSTRALIA: THE BOMBING OF DARWIN, 1942

Amongst the exhibits in the Australian War Memorial is a beige cotton working dress with nine metal buttons, each bearing the insignia of the Australian Military Forces.[18] Manufactured by David Jones, the pleated dress has a smart collar and short cap sleeves edged with narrow, now faded, cerise bands. Above the large, practical front pockets and loose pleats is an original hand-sewn repair, marking the site where its bearer sustained serious shrapnel injuries on 19 February 1942. That morning, Japan bombed Australia.

Lieutenant Joan Somerville

Lieutenant Joan Holland Somerville joined the Army as a 25-year-old physiotherapist on 17 September 1940, and was posted to 2/1 Hospital

---

[16] Spence, J., 'Bangka Island', World War II 1939–1945, available at: http://www.anzacday.org.au/history/ww2/anecdotes/bangka.html Accessed 12 June 2014.

[17] See: 'The Sinking of the *Vyner Brooke*', Australian War Memorial, available at: http://www.awm.gov.au/units/event_302.asp Accessed 16 June 2014.

[18] Australian War Memorial REL/17560 Working overall dress: Lieutenant J. H. Somerville, Australian Army Physiotherapy Service.

Allied health care and the struggle for recognition

Working overall dress worn by Lieutenant Joan Somerville, Australian Army Physiotherapy Service. The dress is on display in the World War II Gallery at the Australian War Memorial, Canberra. *Image courtesy of the Australian War Memorial*, REL/17560.

Ship *Manunda*. She had undertaken her physiotherapy training in Sydney, where her father, Colonel George Cattell Somerville, was deputy director of recruiting, Eastern Command, 2nd AIF.[19] Between November 1940, when Joan was first attached to the 2/1 Hospital ships company, and September 1941, the *Manunda* made four trips to the Middle East.

Sailing from Darling Harbour, the *Manunda* arrived in Darwin on 14 January 1942. Over the next five weeks the medical staff visited military hospitals in the Darwin area and watched the build-up of defensive troops as Australia responded to the Japanese advance through the Indonesian archipelago. Australian forces were reinforced by US ships and P40 Kittyhawk squadrons at Darwin airfield. For those on the *Manunda*, the days were filled with evacuation drills, training of nursing orderlies and adjusting to the routines of life on board the 9115-ton steamer.

---

[19] J. K. Haken, 'Somerville, George Cattell (1877–1959)', *Australian Dictionary of Biography*, National Centre of Biography, Australian National University, http://adb.anu.edu.au/biography/somerville-george-cattell-8580/text14979 published in hardcopy 1990, Accessed online 12 June 2014.

As anxieties about Japan increased, the Australian War Cabinet ordered the evacuation of all 'non-essential' people from Darwin. Although more than 3,000 civilian personnel remained, most of the women, children and infirm had left prior to Christmas. On 15 February, Darwin received news of the fall of Singapore. The war was becoming perilously close to Australia's northern coastline.

At 0930 on 19 February 1942, Lieutenant Somerville, along with the remainder of the medical and nursing staff, were on board the *Manunda* as the first reports of a Japanese approach were received by Darwin radio.

As a flight of Japanese Zero fighters crossed the Timor Sea, the *Manunda* sat in Darwin Harbour. It was virtually undefended and dangerously close to two other vessels, the *Neptuna*, packed with explosives, and the *USS Peary*, a destroyer and primary Japanese target.

Within minutes explosives and incendiary bombs hailed down on the decks on the hapless fleet and churned the Darwin waters into a whirlpool. The trapped ships still at anchor were unable to escape. A dense, foul-smelling black cloud drifted across the harbour and the sea became thick with oil and debris. Hammering anti-aircraft batteries did little to deter the Japanese, many of whom were veterans of Pearl Harbor. The sky darkened with the dense wall of black smoke rising from the harbour.

Surrounded by burning ships, and listing herself, as a result of a 'near miss' aimed at the adjacent *USS Peary*, the decks of the *Manunda* were sprayed with shrapnel, killing four people.

The *Manunda* was the second-largest ship in harbour and clearly marked as a hospital ship. Despite her visible red crosses,[20] a single Zero flew in at low altitude, with its bomb bays open, dropping a well-aimed incendiary into the aft hatch:

> The bomb went through the skylight of a lounge that was used
> as a music room on 'B' Deck and turned the room into a mass
> of twisted debris. All that remained of a grand piano were a few

---

[20] Commander Fuchida denied any intention to attack either the hospital ship *Manunda* or the RAAF hospital on shore (which was completely destroyed). He blamed the attack on the pilot and insisted that he had disobeyed orders, see: Hall, Timothy, *Darwin 1942: Australia's Darkest Hour*, Methuen Australia 1980, Methuen Sydney, p. 43.

strands of twisted wire, and the steel bulkheads and deck were buckled as though made of tin.[21]

By this time fire had broken out, the medical and nursing staff quarters being totally destroyed. Men were flung overboard by the explosion and some of the life-boats were manned by the hospital crew to rescue injured men from the water.[22]

Not satisfied, the pilot returned, strafing the deck with machine-gun fire and shrapnel. Eleven members of the ship's crew were killed on the *Manunda* and 18 seriously wounded. There were also many other minor wounds.[23]

Joan was critically injured. She sustained multiple shrapnel wounds that tore through her working dress. Her 26-year-old friend, Sister Margaret de Mestre, who had served with Joan in Egypt, was not so fortunate. Hit in the back by shrapnel, Margaret died of her wounds two hours later.

Although her navigational equipment was wrecked, the *Manunda*'s main engines were undamaged and she was able to proceed to sea. Despite the damage, she continued to function as a hospital ship:

> The lifts between the decks were no longer functioning, but patients could be carried up and down the narrow companion ways. At the aft end of the ship surgeons worked continuously to treat the injured and the dying. The white hulled ship, conspicuous with its large red crosses on the funnels and deck, became a floating casualty clearing station, as the wounded were brought to her side, many of them suffering from terrible burns from the burning oil.[24]

Eight days later, *Manunda* limped into Fremantle harbour to deliver the wounded.

---

[21] Hall, Timothy, *Darwin 1942: Australia's Darkest Hour*, Methuen Australia 1980, Sydney, p. 42.
[22] 'HMS Manunda History', available at: http://www.far-eastern-heroes.org.uk/Love_Sprang_From_Batu_Lintang/html/ahs_manunda_history.htm Accessed 12 June 2014.
[23] ibid.
[24] Hall, T., p. 43.

Nurses and physiotherapists aboard the Australian hospital ship *Manunda*, c. 1941.
Left to right: Alyson Mills, Joan Somerville, Margaret De Mestre, Lorraine Blow.
Lieutenant Somerville was critically injured and Sister De Mestre was killed in action when
the *Manunda* was bombed by the Japanese on 19 February 1942.
Image courtesy of the Australian War Memorial, donated by J. McKillop. P01081.005.

Joan Somerville underwent surgery for her wounds, but did not return to the ship. She was medically discharged from service on 1 March 1942. In a vagary of recordkeeping, her discharge authority was misplaced. As "no trace could be found of the discharge ever having been recommended, approved or effected, "it was assumed that she 'just walked out.'"[25] Such a suggestion was particularly galling given her status as 'wounded in action' and her family's strong military pedigree.[26]

---

[25] Service records NX 70312 Masseuse McKellop, J.H. (née Somerville) – discharge dated 30 Nov 44 SA/SA2/EB.

[26] Joan's father, Colonel George Cattell Somerville, served with distinction in both World Wars. Wounded at Gallipoli, his subsequent war service in Europe was recognized with the award of the Distinguished Service Order and the Belgian Croix de Guerre. He was appointed CMG – Commander of St Michael and St George – and mentioned in dispatches five times. See: J. K. Haken, 'Somerville, George Cattell (1877–1959)', *Australian Dictionary of Biography*, National Centre of Biography, Australian National University, http://adb.anu.edu.au/biography/somerville-george-cattell-8580/text14979 published in hardcopy 1990, Accessed online 7 June 2014.

Allied health care and the struggle for recognition

Captain Alison McArthur Campbell

In 1944, the newly named Australian Physiotherapy Association was successful in securing the appointment of a Chief Physiotherapist to the Army. The task was given to a woman: Captain Alison McArthur Campbell. Alison had signed on as one of the first serving female physiotherapists in 1940 and served in Egypt, Libya and Palestine with the 2/4th AGH (see above). She had personal experience of physiotherapists' roles and duties in wartime; she also had first-hand experience of the conflicts that came with unclear chains of command, ambiguous authority and professional accountability. Her service diary entries and collection of poetry, written whilst deployed, refer clearly to the strained relationship between the physiotherapists and their matron on one hand, and the medical officers they reported to for their professional duties on the other. In a poem titled 'Syrian Campaign', written in July 1941, she writes:

> Here there is work for all to share
> Heat and sweat and toil and flies
> Fever and the stench of wounds
> Plaster, dust and dysentries
> Suffering is all around
> Together we can ease some pain
> But no unity is found
> Strife and friction seem to reign
> Jealousy the green of eye
> Pettiness that spoils a team
> God! There is so much to do
> This is just a dreadful dream[27]

Under direct control of the Director of Surgery, Alison began a process of reform, promotions and reorganisation of the Army's fledgling physiotherapy service.[28] A seniority list was compiled and 26 of the senior

---

[27] Alison A McArthur Campbell notebook of poems written while on overseas service in AWM – PR90/022.
[28] ibid.

members were promoted to the rank of Captain.[29] Her job included acting as liaison officer between the DGMS and all physiotherapists working in Australian general hospitals and convalescent depots. She also inspected the work of physiotherapists employed in medical units and kept a check on the adequacy and suitability of equipment used in Army physiotherapy departments. She was a pioneer who many would follow.

Pharmacy

Getting medical supplies to frontline medical units in times of war, often across vast distances, with issues of temperature, storage and risk of damage, is immensely challenging. The responsibility falls to Army pharmacy, one of the fundamental disciplines within the profession of military health.

Military pharmacy differs from civilian and hospital pharmacy and has emerged as a separate stream within the Medical Corps. In times of war, production of drugs and surgical instruments takes on critical importance and must be managed alongside supply and dispensing. Large bulk orders and military schedules are on a scale unfamiliar to community pharmacists. The importance of ensuring adequate medical supplies – including medicines, compounding lotions, creams, capsules and suppositories – to overseas troops cannot be underestimated.

During World War II, chemicals were in short supply, particularly alcohol, which was used in vast quantities in hospitals.[30] The United States and Britain both suffered critical shortages of key drugs, that became scarce in the early 1940's. Britain suffered shortages of insulin when its manufacturing plants were destroyed by bombing; the US introduced a voluntary recall of all quinine-containing compounds and there were key drug substitutions to ensure adequate military supplies of essential ingredients. World supplies of quinine, (the only effective antimalarial agent available in 1942) came primarily from Java, now under Japanese control. This scarcity of quinine increased the importance of the work undertaken by Australia's malaria-research teams, led by Major Josephine

---

[29] Walker, A.S., p. 422.

[30] Ethanol is also a critical component in the manufacture of 'smokeless gunpowder'.

Mackerras and others. Their work led to the introduction of the synthetic antimalarial Atebrin (mepacrine).[31]

Penicillin, one of the greatest discoveries of the war, and one that would change the tide of history, was also in short supply.[32] Very limited quantities of penicillin were being produced in England and America, contributing to the decision for CSL (Commonwealth Serum Laboratories) in Melbourne to attempt to produce penicillin on a large scale.[33] New formulations of penicillin – ear-drops and penicillin cream – were also introduced in Australia.[34]

Australian women, particularly those with an education in science, participated in the manufacture of drugs. The Australian Army was the first to adopt the antibacterial sulphaguianidine (a derivative of the early agent sulfanilamide) used with great success to treat the omnipresent dysentery during the Papuan campaigns.[35] These 'sulpha drugs' were produced by Monsanto Australia and credited with a significant impact on the Kokoda Campaign.[36] Vitamin products were also made in Australia during the war, produced from livers of snapper and sharks by the Colonial Sugar Refining Company. Morphine was also in short supply, leading to the development of a new process in Australia whereby opium concentrate was made by extracting opium from the poppy capsules and from the stalks. Work was carried out on antiseptics and the manufacture of new insect repellants, effective against mosquitoes and scrub mites.

Other women took part in the manufacture of surgical instruments. Mass-produced hypodermic syringes were sharpened by hand and adhesive

---

[31] See chapter 3.

[32] Howard Florey, an Australian working at Oxford, was responsible for developing penicillin and for mass production, which enabled its widespread use during World War II. Florey attended Adelaide Medical School and was a Rhodes Scholar.

[33] The Women's Land Army was engaged to harvest pyrthrum at Armidale – a project that proved unsuccessful and was later abandoned. See Walker, Allan, S., 'The Island Campaigns', Australia in the war of 1939–1945, Series 5 – Medical, Volume III, ed. Allan S. Walker et al, Australian War Memorial, 1957, p. 261.

[34] This latter is a success of Fauldings in Australia.

[35] Brewer, A. E., 'The use of suphaguianidine in bacillary dysentery', *BMJ*, 4 Jan 1943, pp. 36–40.

[36] ibid.

strapping was manufactured by gluing rubber on to cotton bandages with gelatin. Women, including those with children, were encouraged to work and contribute to the 'national duty'.

Female pharmacists in the Australian Army

Although women were increasingly joining the workforce to support the war effort, there were relatively few female pharmacists in civilian practice during World War II, even though they had received professional recognition since the late nineteenth century. In 1882, Sarah Ann George (1839–1919) was one of the first women to formally register as a pharmacist in the state of Victoria.[37] As manpower shortages worsened through the war years, women were increasingly encouraged to fill the vacancies created by male pharmacists serving overseas – both in community pharmacies and in military hospitals.

In the 1940s a small number of female pharmacists worked with the Australian Army Medical Women's Service (AAMWS) in dispensaries. As pharmacy technicians, they undertook important roles in the maintenance of drug stocks, inventory management, assisting pharmacists and recordkeeping. Among these women was Staff Sergeant D. Flemming of Melbourne, a qualified chemist whose husband, an A.I.F. Major, was captured during the fall of Singapore and became a prisoner of war.

For the most part, however, during World War II the Army preferred to use nursing sisters as dispensers rather than specifically trained pharmacists. One of these women was Nell Hannah, who preferred to be known as Mavis in her early years. Mavis studied pharmacy and trained at the Royal Adelaide Hospital in the Blood Transfusion unit before the war. Enlisting as a nurse, she was among those who escaped Singapore, just before it fell to the Japanese. Following the sinking of the SS *Vyner Brook*, Nell was one of the Australian nurses who survived internment by the Japanese until her release in September 1945.

---

[37] 'George, Sarah Ann (1839–1919)', The Australian Women's Register, available at: http://www.womenaustralia.info/biogs/PR00756b.htm Accessed 12 June 2014.

Allied health care and the struggle for recognition

Corporal M. Patterson, Australian Army Medical Women's Service (AAMWS), works in the dispensary at 115th Australian General Hospital. *Image courtesy of the Australian War Memorial*, 100370.

Lieutenant Gwyneth Richardson

Gwyneth Jane Richardson was the first woman pharmacist enlisted into the Australian Defence Force. She was commissioned as a Lieutenant on 4 August 1944, after four years of service as a civilian and as a non commissioned officer in the AAMC.

Gwyneth trained as a pharmacist at her family's pharmacy in Annerley, a suburb of Brisbane. She graduated in 1940 and registered with the Pharmacy Board of Queensland on 2 January 1941 (Registration no. 1326). She initially worked for the Army as a civilian pharmacist.

After the Japanese attack on Pearl Harbour, Australia was drawn into the Pacific War and gender-based restrictions on military recruitment were, of necessity, abandoned. Gwyneth Richardson applied to join the defence force and enlisted into the Australian Army Medical Womens' Service (AAMWS) on 11 December 1942. She was 25 and went on to serve in uniform for the next five years.[38]

---

[38] Service Record, QX57786 Richardson, Gwyneth Jane. Australian War Memorial

Her initial rank was staff sergeant and she was paid 6/8d per day.[39] After commissioning as a Lieutenant in 1944 she was appointed to the 2 Australian Women's Hospital based in the gracious heritage manor *Rhyndarra* [40] on the banks of the Brisbane River at Yeronga. She served there until 1946 when she was posted to Port Moresby and Rabaul, at the northern tip of New Britain Island. She worked there as a military pharmacist until July 1946, several months after the end of World War II.

Gwyneth embarked for Port Moresby on 12 March 1946. It was an anxious time for the diminutive young woman, with blue grey eyes. Although Papua was considered safe for women in the final days of the war, disturbing rumours of Japanese atrocities against prisoners and Europeans were proving to be well-founded. Stories were emerging from the prisoner of war camps in Sumatra and Singapore; the murder of two nurses, Mavis Parkinson and Sister Frances Hayman, at a mission in Papua was also horrifying news.[41]

March is typically the wettest month in Papua New Guinea and the daytime temperatures rarely fall below 32 degrees C, resulting in sweltering humidity. Like the women of the AAMWS who were posted to Bougainville, Gwyneth's day dress was a more practical tropical uniform of light-weight khaki and ankle socks instead of stockings. After 6pm regulations mandated an 'anti-malarial' rig of safari jacket, trousers, woollen socks and boots. Lipstick was considered *de rigueur* for women, but needed to be applied frequently and stored in a cool place to avoid melting.[42]

Conditions after the Kokoda campaign were particularly tough. It had been a ferocious jungle campaign, fought on our northern doorstep, and although it ended in January 1943, the conditions were still trying for those

---

[39] ibid.

[40] Rhyndarra was built in 1889 by the Italian architect Andrea Stombuco. In 1942 it was requisitioned by the Australian Army to form part of the 2 Women's Hospital. After the war it remained under Army control as 1 Military Hospital until its closure in 1996.

[41] International Military Tribunal for the Far East proceedings published 20 December 1946 (Webb War Crimes Commission). This commission heard evidence from Vivian Bullwinkel about the Bangka Island massacre and also evidence from Aitere, a local Papuan who gave eyewitness evidence about the bayoneting of Parkinson and Hayman.

[42] Adams-Smith, Patsy, *Australian Women at War*, p. 198.

who remained. Papua New Guinea remains one of the most remote and rugged areas in the Asia-Pacific and many of the soldiers passing through the medical facilities – even the replacement soldiers – were suffering the effects of malaria, tropical heat stress and chronic ulcers. Gwyneth's days were filled with unpacking supplies and arranging distribution to remote detachments, dispensing drugs and cross checking medications before they left the dispensary.

The unconditional surrender of Germany in May 1945 and Japan in August 1945 meant an end to hostilities and the prospect of a return to civilian life. Yet demobilisation meant large numbers of veterans returning home with injuries and illnesses acquired during their service. The demands on pharmacists continued as many servicemen required ongoing treatment for malaria and other conditions.

On her return from Papua New Guinea Gwyneth Richardson was posted to 118 Australian General Hospital, before taking her final discharge on 1 April 1947. By coincidence, she left on her unit's birthday.[43]

BRITISH COMMONWEALTH OCCUPATION FORCE – JAPAN (1946–1951)

Lieutenant Elizabeth Wunsch

After the end of the Pacific War another pharmacist, Lieutenant Elizabeth (Bessie) Wunsch, spent 14 months, in 1950–1, as the senior pharmacist at the Camp Dressing Station, Ebisu, Tokyo.

From 1946 to 1952 Australian forces were responsible for the military occupation of Hiroshima Prefecture, the site of the first atomic bomb attack in history. During this time Australia's role changed from 'occupying power' to 'protective power' in Japan. In 1950, Australian forces were redeployed from Japan, under UN command, to operations in Korea.[44]

---

[43] Pharmacy History Australia: The newsletter of the Australian Academy of the History of Pharmacy, Nov 2004 (vol 2. No. 24).

[44] Wood James, The forgotten force: the Australian military contribution to the occupation of Japan 1945–1952, Allen and Unwin, 1998 See also: Wood J. The Australian Military Contribution To The Occupation Of Japan, 1945–AWM 114 130/1/42 Japan.

At its peak, the British Commonwealth Occupation Force (BCOF) comprised 12,000 personnel. Of these 92 were women, primarily serving with 130 Australian General Hospital. 130 AGH was located on Eta Jima, where female members were accommodated in dormitories of 22, living under the extraordinarily stringent restrictions placed on the movement and activities of all Australian women serving with the BCOF. Given the treatment of Australian nurses as PoWs and the unequal rights of Japanese women, they did not mix socially with the Japanese.

Elizabeth had qualified as a pharmacist in 1929. Previously a chemist's assistant in Wagga, she passed her pharmacy examinations (with credit) in December that year. She enlisted into the AAMS in March 1943 and served at the No. 3 Australian Servicewomen's Hospital (3AWH) located at Concord in the suburbs of Sydney, as a Staff Sergeant pharmacist.[45] Here, in spacious grounds adjacent to 113 AGH, the establishment of a servicewomen's hospital brought New South Wales into line with Victoria and Queensland by providing dedicated health facilities for women of the Defence Force.

On 31 October 1949 Elizabeth was promoted to Lieutenant and "APPOINTMENT TO RAAMC (PHARM) INTERIM ARMY IN RANK LT FOR POSTING AS PHARM DCS TOYKO TO DATE 21 OCT 49 (.)".[46] It is not clear what her duties as a pharmacist with the BCOF were, but she is quoted in the *Sydney Morning Herald*, saying:

> We did not get to know very much about the Japanese
> pharmacists. The quality and standard of the drugs they use, and
> the supply, did not appear to be as high as it is in Australia, and
> the latest antibiotic drugs are still fairly scarce.[47]

She also explained, for the same article, that many doctors in Japan did their own dispensing, and only a relatively small amount went through pharmacists. Almost certainly, like Army pharmacists in the current ADF,

---

[45] *Sydney Morning Herald* (NSW: 1842–1954), Thursday 1 April 1943, p. 3.

[46] M33(c)/IV 241515 K Military Secretary Army Melbourne to MILCOMMAND Sydney dated 24 Oct 49.

[47] *Sydney Morning Herald* (NSW: 1842–1954), Thursday 11 January 1951, p. 11.

Elizabeth's role would have included inventory management and re-supply, in addition to dispensing and compounding responsibilities.

Lieutenant Wunsch returned from Japan in late 1950 and was discharged on 15 January 1951, aged 44 years.

Dietetics

The establishment of large military hospitals during World War II required a new workforce, including dietitians and occupational therapists.

Prior to World War II, the Army had not employed dietitians and the profession was still in its infancy. Initially, the Australian Army appointed qualified dietitians to its base hospitals at Brisbane, Sydney, Melbourne, Adelaide and Perth. Originally brought in as Specialist Officers of the Voluntary Aid Detachments, this proved unsatisfactory as they held no official rank or status. This created difficulties in a military environment and was inconsistent with the conditions of service for other 'scientific' women.

During the war, 12 dietitians were appointed on fulltime duty as Lieutenants (with provision for promotion to Captain and Major) within the AAMWS, and subsequently transferred to the AAMC. Captain Alice Wunderly was appointed to coordinate with the Australian Army Catering Corps regarding the feeding of patients and nutrition on the wards.[48]

Dietetic roles included catering management in each of the hospitals, as well as the provision of clinical dietetic care for a wide range of conditions. Many of the diseases and injuries experienced by military personnel were very different from those usually encountered in civilian hospitals. Acute-care patients included plastic surgery cases such as soldiers who had suffered gunshot injuries to the head and face and required extensive facio-maxillary surgery. Often their jaws were wired together, and the only way to feed them was through a tube using mixtures of fortified milk, liquid meat and vegetables and fruits.[49] The provision of heated food trolleys

---

[48] Walker, Allan, S., 'The Island Campaigns', Australia in the war of 1939–1945, Series 5 – Medical, Volume III, ed. Allan S. Walker et al, Australian War Memorial, 1957.

[49] Turner-Cats, C. N., 'Retrospective. Dietetics in Wartime', *Australian Journal of Nutrition and Dietetics*, 1993:50(4): p. 174–176.

in hospital wards and special soft diets for patients with facial and oral injuries were advances that made a significant difference to outcomes.

The dietitians' workloads were substantial. In the later years of the war military hospitals such as Concord in New South Wales, Heidelberg in Victoria, Daw Park in South Australia, Hollywood in Western Australia and Holland Park and Greenslopes in Queensland had many ill, injured and convalescing soldiers, many of whom were suffering from malnutrition.[50]

Captain Caroline Turner

The experiences of Caroline (Nancy) Turner, who served as a dietitian during World War II, are typical of others in her profession.[51] Caroline graduated from Melbourne University in 1939 with the degree of Bachelor of Science and the following year (1940) with a post-graduate Diploma of Dietetics.

At the outbreak of war she was working as a dietitian-housekeeper at the 120-bed Castlemaine and District Community Hospital. She worked with the Red Cross conducting evening courses for volunteers to train them in nutrition and large scale cookery. In January 1942 she enlisted. Due to ongoing issues regarding women's employment in the Army, dietitians were enlisted as VAD Specialist Officers. To distinguish them, they wore a purple stripe on their shoulder tabs, which caused some confusion:

> We were often mistaken for 'returned wounded', as 'returned wounded' men wore a pale blue shoulder stripe. Our outdoor uniform was navy blue overcoat and suit (tailor made for officers) with red cross and colour patch, white blouse and striped tie, and navy hat with the 'rising sun' badge. Our underclothing was army issue and I recall my embarrassment when having to go to the quartermaster's store for replacements and having my oldest underwear held up for inspection by all and sundry in the store.[52]

---

[50] ibid.
[51] Roberts, N., 'Obituary: Caroline Nancy Cats', *Australian Journal of Nutrition and Dietetics*, 1997:54(4):p. 206–8.
[52] Turner-Cats, C.N.

Allied health care and the struggle for recognition

Later, when dietitians were commissioned into the AAMWS, Caroline was issued with khaki, captains pips for her epaulettes and a red cross. Later still, she was moved to the AAAMC – in all she was transferred between three different services in four years.

Although many of the dietitians had volunteered to serve overseas, none were sent out of Australia. Instead, most were posted to large base hospitals, where they were responsible for the nutrition of the patients and for teaching Warrant Officers the finer points of running a diet kitchen.

Major Audrey Cahn, Captain Caroline Turner and two lieutenants were posted to 115 Heidelberg Military Hospital (HMH); a large pavilion hospital erected on the outskirts of Heidelberg. Here, they were responsible for two diet kitchens and the convalescent messes (one for officers and one for other ranks). As the war progressed rationing was introduced, and certain foodstuffs such as sugar, butter and prunes were difficult to obtain. Caroline recalled that:

> Food was issued according to ration scales with special allowances to hospitals and some original ideas from 'nutritional experts' at headquarters. At one time, we were sent dried peas with orders to sprout them and use them as a good source of vitamin C. The men refused to eat them because the sprouts looked like maggots.[53]

Nutrition for the returning prisoners of war was a particular challenge. After V-J Day, Captain Turner was sent north to No. 102 Holland Park General Hospital in Brisbane, where she was the sole dietitian in charge of five kitchens, two diet kitchens, six messes, two WO catering officers and two WO AAMWS diet supervisors. As the first staging hospital for returning prisoners from Malaya and the Pacific Islands, convoys arrived regularly by hospital ship bringing emaciated and broken men who had been kept under terrible conditions in prisoner of war camps by the Japanese. The men had suffered prolonged starvation, vitamin deficiencies and tropical diseases. Years of malnutrition had rendered their stomachs incapable of coping with normal quantities of food so adjustment was gradual. They began with small, frequent meals of soft, high-nutrient

---

[53] ibid.

feeds.[54] The management of 're-feeding syndrome' required precise and measured introduction of nutrients to prevent complications that led to burning swollen feet, heart disorders, neurological problems or in some cases, death. Synthetic vitamins were not available, so all concentrates had to be hand prepared. For many of the men, being encouraged to eat by the dieticians and ward staff was a vital part of their psychological, as well as their physical, recovery.

As the men became well enough to travel south by troop train they would depart the hospital for home. In mid-1946, after four and a half years in service, Captain Caroline Turner returned south to Melbourne, was 'demobbed', and returned to civilian life. Thirty-five years after war's end, she married Volkert Cats – a Dutch airman she had first met in Melbourne during the war in 1942 when he was recovering from experiences against the Japanese in Java[55].

OCCUPATIONAL THERAPY

By 1941 the role of occupational therapy in military hospitals was gradually maturing. At this time, only three Australian women held diplomas in occupational therapy – Misses Keam, Francis and Docker – who were tasked with training technical assistants to fill vacancies in base hospitals. Usually recruited through the VAD organisation, the assistants received six to nine months training in occupational therapy. After graduation they were commissioned as Lieutenants in the AAMWS. The first appointment took place in April 1944 and by the end of the war there were 30 trained occupational therapists in the Army.

Occupational therapists were appointed to static hospitals only. Ultimately, they were transferred from the AAMWS to the AAMC, though on 1 September 1944, 16 occupational therapists on full-time duty with commissioned rank were transferred from the AAMWS to the AAMS. Working closely with physiotherapists, orthopaedic surgeons and

---

[54] Schnitker, M. A., Mattman, P. E., Bliss, T. L., 'A clinical study of malnutrition in Japanese prisoners of war', *Annals of Internal Medicine* 1951; 35: 69–96.
[55] Roberts N.

Red Cross volunteers, these women made a significant contribution under the stewardship of Major J. M. Keam.

A large number of women also served as administrative staff, stores staff, drivers, cooks, electricians and hygiene personnel in medical units. Many never wore RAAMC embellishments and their service is often overlooked. Nonetheless, while not contributing directly to patient care, these women were 'silent facilitators' and should be acknowledged.

The VAD and the AAMWS – forerunners of the army medic

Much of the womanpower for the newly formed AAMWS was derived from the Volunteer Aid Detachments (VADs). The VAD service had lapsed during the inter-war years, but was again called into service in 1939. Almost immediately VADs began to replace men in military hospitals, undertaking duties as medical orderlies and nursing assistants on a voluntary basis.[56]

Although they were often regarded simply as 'nursing assistants', the VADs and the VAD Specialist Officers undertook a wide range of activities. Many years later, their roles would merge into one – the 'Army Medic'. For now, however, the service of the humble VAD was mired in discrepancy.

Initially, paying VADs was thought to prejudice paid employment for men, but by 1940 the War Cabinet decreed that they should receive remuneration at the same level of a private soldier.[57] Miss B. Foll was the first full time Voluntary Aid to be employed in a military hospital on a paid basis and commenced her duties at the Showground in Sydney on 1 January 1941.[58] At the same time recruitment occurred elsewhere in the country, at Heidelberg in Victoria and Enoggera in Queensland. To be acceptable applicants had to be unmarried, without dependants, between the age of 21 and 45 years, and willing to work hard.

---

[56] Walker, A.S., chapter 37.

[57] ibid.

[58] 'Voluntary Aid Detachments', *Vetaffairs*, DVA, Summer 2011, p. 3.

Not For Glory

Recruiting poster for the Women's Army Corps.

Once again pay rates, status and service discrepancies became a source of conflict. Nurses were aghast that untrained VADs received the same pay and allowances. Neither group wore any army rank, but nurses were considered as holding 'officer status' while VADs were regarded as 'other ranks'. Disputes over legal status, veteran entitlements and enlistment procedures further complicated the issue.

There were also complaints about the term 'medical women' (as in AAMWS), with some arguing that it implied women with medical degrees. The Secretary of the Department of the Army thought differently, arguing that any confusion was unlikely. The VADs meanwhile, along with other members of the AAMWS, were caught in a 'no man's land' – they were neither nurses nor medical orderlies. Plagued not only by a lack of clarity about their role, their pay and their status, many nurses resented the VADs serving overseas and saw them as 'earnest sparetime amateurs'. Despite this, they performed valuable work as auxiliaries, drivers, laboratory assistants and radiology assistants – roles that would all subsequently come under the domain of the RAAMC.[59]

Glynneath Cody (née Powell)

At the outbreak of World War II, Glynneath Cody was one of many women who signed on as a VAD.[60] Like many of her colleagues, she served in Australia and did not go overseas. Given the same blue uniform used by

---

[59] *Paulatim*, Magazine of the Royal Australian Army Medical Corps. Special AAMC Centenary Edition 2003. p.381.

[60] Another of the 'wellknown' VAD/AAMWS was Joan Richardson, also from Perth. Her story is included in Patsy Adams Smith's *Australian Women at War*.

Allied health care and the struggle for recognition

Studio portrait of Sergeant Glynneath L. Powell (*née* Cody), Australian Army Medical Women's Service (AAMWS). *Image courtesy of the Australian War Memorial, donated by G. Cody.* P01985.002.

VADs during World War I, she received one month's training, juggling her new VAD duties with part time work as a bookkeeper.

Born in Western Australia in 1917, three of her uncles served at Gallipoli and two later died on the Western Front. Her father died when she was aged 14, cutting her education at Boulder Primary School in the goldfields short. Instead, she took up paid work (a difficult task during the Depression years) as a cashier in a department store. Determined to continue her education, Glynneath attended night school four evenings each week.

By mid-1942, Glynneath's remaining uncle was killed at Tobruk, during his service with the 2nd AIF. Glynneath's VAD duties began at Northam – a two-ward Camp Hospital with one medical officer and four nursing sisters. Her work was typical: taking temperatures, making beds, sponging patients and assisting the nurses. On occasion she would be permitted to help with inoculations or dressing wounds.

Glynneath quickly demonstrated her potential and was given responsibility for training other VADs. By the time the VAD service was subsumed under the AAMWS she had risen to the rank of Sergeant. Like her other VAD colleagues, many of whom felt a passionate attachment to their blue uniforms and distinctive red cross chevrolets, Glynneath was sad to transition to the more practical khaki of the AAMWS.

When Northam Camp Hospital closed, Glynneath was transferred to the much larger Hollywood Hospital, based in Perth. As was the case in similar hospitals elsewhere, the AAMWS members lived in huts of 20 women with each space divided by wardrobes into cubicles. They worked long hours, either on the wards or as general duties orderlies – helping out in kitchens, messes and pantries. For some, other opportunities arose and they were assigned to help in specialist areas, such as the radiology or pathology departments. Glynneath spent 18 months working in the blood bank at Hollywood – the same blood bank that Lieutenant Jean Kahan had helped to establish.

Despite their long shifts and six-day working weeks, life was not entirely dull for the women of the AAMWS. American and Australian soldiers were keen for female company and would exchange coupons – for tea, coffee or chocolate. Although she continued to serve until after the end of the war, Glynneath met her future husband during her time at Hollywood. Les Cody had served with the 2/4th Machine Gun Unit and was repatriated after being a prisoner of the Japanese on the Thai-Burma Railway. The couple married in 1948.

In 1993, Glynneath travelled to Canberra to carry the AAMWS banner at the entombment of the unknown soldier and witnessed the then Prime Minister Paul Keating's eulogy: words that have since been inscribed on the tomb.[61]

---

[61] The eulogy was delivered by then Prime Minister Paul Keating on 11 November 1993 at the entombing of the Unknown Australian Soldier and included the following words: *We do not know this Australian's name and we never will. We do not know his rank or battalion. We do not know where he was born, nor precisely how he died ... We will never know who this Australian was ... he was one of the 45,000 Australians who died on the Western Front ... one of the 60,000 Australians who died on foreign soil. One of the 100,000 Australians who died in wars this century. He is all of them. And he is one of us.*

VADs AND OVERSEAS SERVICE

For some members of the AAMWS, their service took them to remote parts of Australia. A number were posted to medical units in the Northern Territory – at Alice Springs, Katherine, Adelaide River, Tennant Creek, Larrimah and Mataranka, others were posted to Hall's Creek and Mt Isa. Indigenous women also enlisted.[62] In the north of the country, the wet season and the heat were challenging – for many women it was their first experience of Outback Australia.

As AIF hospitals overseas began to experience manpower difficulties a decision was made to extend the scope of the VADs employment. On 27 July 1941, General Blamey approved the posting of 540 VADs to hospitals in the Middle East. The move freed 292 males to take up other duties (and reflected notions of 'relative employment value' between men and women that prevailed at the time).

Although two VADs – a pathology assistant and a secretary to the matron – were posted to 2/2nd Australian General Hospital in early 1940, it was not until October 1941 that disagreements over service conditions were resolved. A party of VADs then embarked on the hospital ship *Wanganella* for service in the 2/12 AGH in Ceylon under the stewardship of Miss M. M. C. Stephen.

Voluntary aides (by now paid but still without rank) saw service in Gaza, Cairo and Sidon. Most returned to Australia after the entry of Japan into the war and there were no further overseas postings until 1943. Then, members of the AAMWS were sent to New Guinea with the 46th Camp hospital and later with the 2/9th and 2/7th AGHs. In November 1943, 50 members of the AAMWS flew across the Owen Stanley Ranges to Buna,[63] where they helped establish the hospital. Under the supervision of the

---

[62] Kathleen Ruska, of the Noonuccal people of Stradbroke Island, enlisted in the Australian Women's Army Service – she later gained fame as the celebrated poet Kath Walker, ultimately reverting to her Indigenous name, Oodgeroo Noonuccal. Her cousin, Winnie Iselin, joined the Australian Medical Women's Army Service. See, 'A brief history of Indigenous Australians at War' by Moremon, John. Available at: http://www.dva.gov.au/benefitsAndServices/ind/Pages/at_war.aspx Accessed 9 June 2014.

[63] Walker, A.S., chapter 37.

medical officers, they cared for patients and performed a range of hospital duties until the arrival of nursing staff eight days later.

In the concluding days of the war members of the AAMWS were also send to New Britain, Bougainville, Hollandia, Morotai and Borneo. They worked hard in a variety of roles, under difficult circumstances, and endured all the usual hazards of service under operational conditions. In one incident, sixteen VADs had a narrow escape from an aircraft that burst into flames on their return to Australia. After a crash landing, the aircraft blew up only seconds after those on board had made a lucky escape.[64]

At the same time, the scope of VAD work at home expanded and the women were organised and administered under the auspices of the Army Medical Service. Issues of rank and command authority remained problematic. Colonel Spiers, Commanding Officer of 2/4 AGH, echoed widespread anxiety about women stepping into the traditional domain of men:

> I feel very strongly that to replace senior NCOs by females would be detrimental to discipline. Some male orderlies are essential for heavy lifting. I hesitate to picture female NCOs ordering ward orderlies about.[65]

The contribution of the VADs/AAMWS has been conflated and confused with that of nurses. One reason was the appointment of a new 'controller' at the AAMWS – Lieutenant Colonel Kathleen Best OBE. An adroit administrator, Lieutenant Colonel Best had previously been a matron with distinguished nursing service. Her background in nursing, and that of her successors, contributed to confusion about AAMWS roles – in some areas it was seen merely as a subsidiary nursing service, rather than an umbrella organisation comprising women from professional 'scientific' backgrounds and those with non-professional skills.

In February 1951, the AAMWS was disbanded and incorporated into the Nursing Corps. With hindsight, this may seem a sensible and

---

[64] ibid.
[65] Unit War diaries 6 Jun 1943 Col NL Spiers CO 2/4/ AGH Redbank (quoted in R Goodman. A Hospital at War (op cit 438 *Paulatim*)

pragmatic solution, but it created distinctions and hierarchies that were not, necessarily, beneficial. For those in non-nursing roles such as medical and allied health support, they were in limbo once again, and caught between two domains – nursing and the Army Medical Corps.

There is no doubt that members of the AAMWS made a significant contribution to the war effort. By August 1945 there were 4,694 officers and ORs serving in the AAMWS at home, with a further 457 women serving abroad.[66]

THE HOME FRONT AND THE WOMEN'S HEALTH SERVICE

By the end of 1942, approximately 30,000 Australian women were serving as part of Australia's Armed Forces. The Department of Defence responded to their medical needs by raising three Australian Women's Hospitals – unique units that had no parallel in the previous war and have never been reconstituted since.[67]

In the 1940s, it was still deemed appropriate for women to be cared for in hospital separated from men. The concept of 'shared wards' was an anathema and most women in civilian society continued to receive care in 'Women's Hospitals'. While the vast majority of women presented with conditions unrelated to gender, there was a not insignificant rate of gynaecological problems, including unplanned pregnancies and venereal disease that required specialist care, particularly towards the end of the war.

On the 'Home Front' there were also Repatriation Hospitals that specifically treated returned soldiers. These hospitals later came under the Repatriation Commission and Department of Veterans Affairs; they were the backbone of a national effort to care for the thousands who returned scarred by wounds, both seen and unseen, incurred during their service.

---

[66] Grey, J., The Australian Army. Oxford University Press 2001 p. 146.
[67] Williams, L., 'No Man's Land: A History of the 2nd Australian Women's Hospital', 1997, Foreword, p. iii.

Captain Joan Refshauge

Joan Refshauge was typical of the female doctors that served in the Australian Army on the 'Home Front' during the war. Born in the rich farming district of Armadale where her father was headmaster, she completed a diploma of education in 1930 and commenced study in medicine in 1935 (she graduated with a MBBS from the University of Melbourne in 1939).

Joan came from a military family and one of her brothers was Major General Sir William Dudley Refshauge, a medical officer who served with distinction during World War II in the Middle East and Tobruk. He was subsequently appointed Director-General Army Health Services.

Joan suffered from gross astigmatism and wore heavy, thick glasses. She was taken out of school at the age of 12 because she was "too emotional and because the fact that blindness might ensue could not be ignored".[68] Her parents were later persuaded to allow her to complete her education and she was sent to a tutoring college. Within six months Joan had completed all eight subjects for the Intermediate Certificate, including music. She went on to matriculate and study music at the Conservatorium. Her father felt, however, that music was a poor influence on her and stopped her music lessons, insisting that mathematics and science were a "stablising influence". As a science undergraduate at university she approached her father about changing over to medicine and explained that there were 'free places' available, so her fees would be paid. The answer was no: "No woman should pursue such a course."[69]

Joan was not put off. She went on to study medicine and, although she married while a student, she continued to support herself by working as a maths tutor at night and taking out a university loan.

Joan undertook her hospital residency at the Alfred Hospital in Melbourne, but recalled that: "The male graduates objected because I was a woman and the women because I was married. But the unpleasantness

---

[68] Hellstedt, Leone McGregor, 'Australia', Women Physicians of the World, Hemisphere Publishing Corporation, Washington, pp. 333–337.
[69] ibid.

passed."[70] The following year, suffering from morning sickness, Joan had to resign her position at the Queen Victoria Hospital for Women, but was able to work at the Chronic and Incurable Disease Hospital for the remainder of her pregnancy.

It must have been was a worrying time for the new mother. Joan recalled that:

> My husband was in Lae, New Guinea, as a civilian, and there was no word from him after the bombings there. When my son was about three months, I was informed that my husband was presumed alive and a private A.N.G.A.U. [Australian New Guinea Administrative Unit] and that monetary adjustment was necessary, backdated to the first bombing of Rabaul. My mother looked after my son, and I became senior medical officer at Mildura Base Hospital. I spent every weekend off duty in Melbourne. My mother became ill, and I resigned my position only to find I was 'man-powered' and had to await approval and replacement. A mothercraft nurse then looked after my son in Melbourne.[71]

Although Joan's description of events is matter-of-fact, the reality of her family circumstances, separated from her husband and son, would have been hard to bear. Her private papers in Australia's National Library include letters from her son, sent a few years later from his boarding school in Melbourne. In one letter addressed to his grandmother, the word 'mummy' is crossed out – the young boy appears to be confused about who he is writing to: 'Mummy' or 'Marnie', the name of his grandmother.[72]

Once her mother's health improved, Joan enlisted into the Army Medical Corps as a Captain Medical Officer. She was posted to the 2/2 Australian Army Out-Patients Department (A.O.P.D.), where she was responsible for the health of women in the army and allied services stationed in and around Melbourne. Her appointment was terminated in April 1946.

Joan's husband remained in Papua and she joined him in Port Moresby

---

[70] ibid.

[71] ibid.

[72] Papers of Dr Joan Refshauge, 1931–1979, MS 7026, National Library of Australia, box 14.

in 1947. She became a pioneer in public health and an advocate of women's rights. In 1964 she was awarded an O.B.E. and received the Cilento medal for tropical medicine.[73]

Captain Joan Crosier

Joan Crosier enlisted into the Australian Army Medical Corps in October 1942, having spent time as a solo doctor in charge of the Red Cross Blood Bank in Toowoomba.

Initially posted to 7 Australian Camp Hospital at Redbank, she was transferred to the recently established 2 Australian Women's Hospital (2AWH)[74] that was also set up at Redbank and remained open for the rest of the war.[75] The other Women's Hospitals were 1AWH at Claremont in Western Australia and 3AWH at Concord in New South Wales.

At 2AWH she worked closely with her colleague, Captain Margaret (Peg) Mackay, a Melbourne graduate, whose parents were missionaries in Korea. Peg was an outgoing and amusing woman and one of the few female doctors who was married – her husband, John Ray, was a doctor who saw service in Greece with the Australian Army before returning to marry Peg. Peg travelled with him to Brisbane and remained there while he served in New Guinea.

During her time at 2AWH, Joan was in charge of the infectious and venereal disease (VD) patients. She was sometimes called upon to give anaesthetics when the specialist surgeons would visit from the nearby 2/4 AGH.

The wards were busy and by 31 July 1943 the hospital was full to capacity with 132 beds occupied. In January 1944, the DDMS agreed to extend the hospital to 200 beds. The medical staff worked seven days a week with alternate nights on duty.[76] The illnesses of the women were

---

[73] Denoon, Donald, 'Refshauge, Joan Janet Brown (1906–1979)', *Australian Dictionary of Biography*, National Centre of Biography, Australian National University, http://adb.anu.edu.au/biography/refshauge-joan-janet-brown-11497/text20507 Accessed 6 March 2013.

[74] Williams, L., *No Man's Land*.

[75] The commanding officer was Captain Ramsey Warden (VX14457), a Victorian gynaecologist.

[76] Williams, L., *No Better Profession*, p. 159.

much the same as those seen in civilian practice – whooping cough, chest infections, diphtheria, rubella and colitis. While there were few wounded patients, there was a small but notable incidence of 'functional disorders' or neuroses and a number of cases of malingering.

Infectious diseases predominated through 1942 and in January 1943 there was an outbreak of dengue. This was particularly alarming as there was (and remains) no specific treatment for the disease. Measles and rubella were also common. At the time, public health campaigns and vaccination regimes were not commonplace; antibiotics were still in their infancy in clinical practice and isolation wards were routine. Death from infectious disease was not uncommon and in the third quarter of 1943, 2AWH recorded its first death – a case of pneumococcal meningitis.

Although surgery was undertaken at 2AWH – there was eye surgery, appendectomies, and gynaecological procedures – it was the medical cases that occupied most of the beds. In September 1944, a specialist physician, Major Helen Taylor, joined the staff and became OC of the Medical Division. Captain Crosier continued her work with the VD cases – she encountered 11 cases of gonorrhea in the April-June quarter of 1943 alone – and worked closely with Dr Beatrice Warner from the Department of Health and the Venereal Diseases Hospital. During these early years, trials of penicillin combined with sulphonamide were undertaken to attempt to control gonorrhea,[77] but this was not the only sexually transmitted disease to cause concern. *Trichomonas vaginalis*, a protozoan that was resistant to penicillin, became an emerging problem. Treatments were notoriously unsuccessful and included vaginal pessaries soaked with iodine or douches containing hypertonic saline and silver solution. Sadly such diseases led to infertility, a problem that some married women attributed (unfairly) to their husbands taking the antimalarial treatment Atebrin.

Unwanted pregnancies were also a problem. In his quarterly report dated 30 June 1945, the Commanding Officer of 2AWH, Lieutenant Colonel Twohig, remarked on the plight of the unmarried woman who became pregnant while serving in forward areas. "The medical papers

---

[77] Sir Howard Florey visited the hospital on 16th September 1944 and discussed the use of penicillin in combination with sulphonamides, proposing a longer course of treatment than that given in the United States.

broadcast her condition to all *en route*." During the quarter there had been 85 deliveries, 10 miscarriages and three septic abortions.[78]

As servicewomen returned from deployment towards the end of the war, there was a significant influx of acute psychiatric cases at the hospital in its final months. Still, the hospital was 'wound up' and, like other units, a system of points was devised – based on length of service, age and martial status – to determine the order in which personnel should be discharged and returned to 'civvie street'. The only Medical Officer to serve at 2AWH during its entire time, Captain Joan Crosier, finished her war service on 19 November 1946.[79]

Captain Gwen Fleming

Gwen Fleming was among the first wave of women doctors to graduate from Sydney University in time to answer the call to serve during World War II. She was appointed to the rank of Captain and became the Major at Yaralla Military Hospital, Concord where she was Officer Commanding the Medical Company. Her colleague, Dr Margery Scott-Young, was charged with the surgical aspects.[80]

The war challenged Gwen in every way. Her specialty was thoracic medicine, but she faced returning soldiers cringing in foetal spasms of shellshock, some disfigured and limbless. Her own brother, Bobby, of the 2/30 Australian Infantry, would die, along with 2,815 other young Australian men on the Thai Burma railway.

At the end of the war she married a dashing young surgeon (RAAF Flight Lieutenant) from the Flying Doctor service. In 1945, she was one of the first women admitted as a Member (and subsequently Fellow) of the Royal Australasian College of Physicians.[81]

Few of these doctors continued their military careers after the war. There remained a prevailing view that although women doctors could be

---

[78] Williams, L., *No Man's Land*.

[79] Joan Crosier went on to specialise in anaesthetics and received her Fellowship of the Faculty of Anaesthetists of the Royal Australasian College of Surgeons in 1952.

[80] 'Dr Gwen Fleming 1916–2011', Your News, *North Shore Times* 7 February 2011.

[81] 'Mary Gwenyth (Gwen) Fleming', Obituary, *Medical Journal of Australia*, 2011; 194 (6):316.

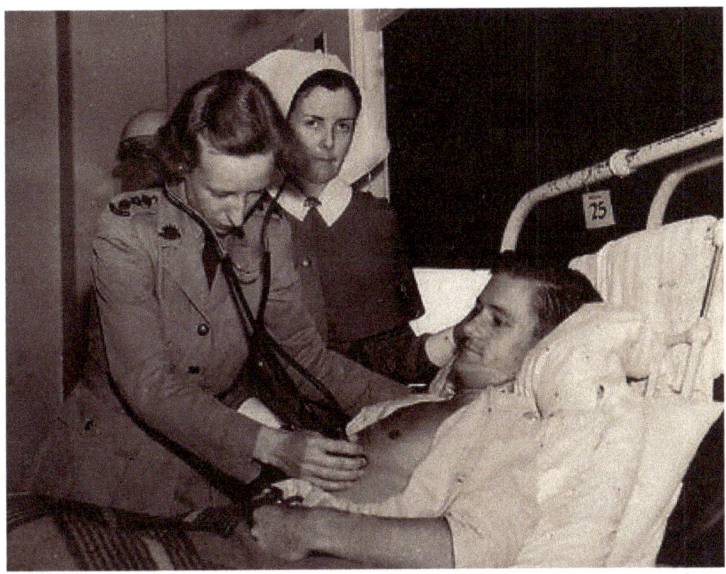

Captain Gwen Fleming examines a patient at Heidelberg Hospital, Victoria.
Image courtesy of her son, Justin Fleming

accepted into the profession, their rightful place was in women's wards and children's hospitals.[82] Like those who served before them in World War I, they quietly hung up their uniforms and returned to civilian life.

AFTER WORLD WAR II

War had demonstrated women's ability in multiple roles. Traditional boundaries had been crossed. Women had worked in factories – previously regarded as a man's domain – and married women had contributed to the war effort, leaving their children at home with carers. The Australian Army had accepted not only married women, but women with children into its officer ranks. Stereotypes had been broken as women served their communities and the country in war.

The future for women's participation in the workforce and the professions looked bright. In a 1946 article published in a supplement to the *Melbourne Argus* newspaper, the achievements of a number of

---

[82] Hutton Neve, *This Mad Folly*, Sydney: Library of Australian History 1980; p. 94.

professional women were recognised, including the then Lieutenant Colonel Lady MacKenzie. Entitled 'Wither Women Next?', the article also acknowledged another Victorian, Dr Meredith Ross, who was Head of the British Girl Guides Association's first medical unit to Western Europe in 1944, and Dame Constance D'Arcy, who was the first woman elected Deputy Chancellor of Sydney University in 1943. The author concluded:

> There is now practically no activity in which women have not been engaged … the recent war found women in all countries doing the hard manual labour usually reserved for men, and there is not a branch of sport wherein they have not distinguished themselves. Is it any wonder I ask: "Wither woman next?"[83]

Still, in postwar Australian society the medical profession remained reluctant to embrace such change. Prestigious hospital appointments that had been filled by women doctors during the war were again allocated to males returning from conflict. Women doctors returned, as they had following World War I, to more 'acceptable' appointments at Women's Hospitals, as gynaecologists, as anaesthetists or in family and maternal practice.

The medical scientific community, too, was slow to adapt. Some women, like Josephine Mackerras, went on to forge significant careers, but for the majority – even though they were 'manpowered' during the war years – there was a return to inequality. Women were often not given full scientific status and were employed in 'non-scientific' roles even though they performed the same research as their male counterparts. They were considered ineligible for scholarships or promotion opportunities. They were also required to retire from work on marriage.[84]

Only one of the doctors who had served during the war remained in

---

[83] Malloch, N.W.

[84] A well-known example is of radiophysicist Ruby Payne-Scott who was involved in Defence research into the use of radar at CSIRO during World War II. Postwar she pioneered work into the field of radioastronomy. She hid her marriage for six years from authorities, rather than retire from her role, but in July 1951 with her first child imminent, she was obliged to resign. See Memorandum A8520, PH/PAY/002 Dr Ian Clunies Ross, Chairman CSIRO to Payne Scott, dated 3 March 1950.

uniform; the widowed Lady MacKenzie. Reverting to the rank of Major, she stayed on as a medical administrator, finally hanging up her uniform in 1946.

Recognition

Recognition of the roles undertaken by medical women during World War II was also lacking in the years that followed. Still, there was an attempt to recognise two of the female army doctors through the Australian Honours system.

Major Lady MacKenzie and Major Josephine Mackerras were each serially recommended for a military decoration (MBE) by Major-General Burston, the Director General of Medical Services.[85] Under a 50-year archives access rule it has become possible to obtain details of these nominations.

In a summary memorandum of 1944 entitled, 'Women Officers – Australian Army Medical Corps'[86] it is stated explicitly that "[women] medical officers receive the same pay and allowances as male officer". However, under the entry specifying the conditions for decorations and awards, a draft proposal that "women officers of the AAMC are eligible for all Honours and Awards available to male officers in the AAMC", has been struck out by the senior reviewing AAMC officer.[87]

Among other summaries of her service, a proposed citation for Major MacKenzie noted her "valuable contribution to the efficiency of the Administration of the Army Medical Services". In the case of Major

---

[85] Australian War Memorial, 'End of War' Awards submission by QMG and DGMS. Rejected Citation (MBE) for Major Mabel Josephine Mackerras NFX 137899 and for Major Winifred Iris MacKenzie VFX 81148, AWM 119-248/25.

[86] This document is contained within the Tait collection of the Australian War Memorial :88/1/1–23. There are 23 folders from a range of sources. There are five files labeled 'Dr Walker's' that were used as background documentation for the Official History, but are not necessarily official documents. One of these files is the original that has the medals notation crossed out.

[87] 'Women Officers – Australian Army Medical Corps. (This memorandum deals with Women Officers of the AAMC and therefore is not concerned with The Australian Army Nursing Service or The Australian Army Women's Service) 1944', Memorandum, Australian War Memorial Archives, File 422/7/8, War of 1939–45 AWN No. 481/2/23; Indexed 88A/2.

Josephine Mackerras, the proposed wording continues to encapsulate her service:

> Major M J Mackerras, who on enlistment was over age for secondment to the AIF, has been entomologist in the LHQ Medical Research Unit (AIF) since its inauguration. By her fine scientific judgement, technical skill, and constant devotion to duty she has organised and developed a technique for the infection of anopheline mosquitoes and human volunteers on a scale never previously attempted. Latterly urgent requests by the Research Councils of U.K. and U.S.A. for the scientific testing of new anti-malaria drugs have necessitated the rapid infection of hundreds of volunteers, and this would not have been feasible without the scientific achievement of Major Mackerras and her entomological team.
>
> Brigadier J. A. Stinton, V.C., Malariologist to the War Office and Major-General Covell, Malariologist to the Gove of India and S.E.A.C., regard this work as unique.
>
> Few women can have made a greater contribution to the Allied war effort."[88]

Both recommendations, submitted on three occasions – to the King's Birthday Honours List in 1945, New Year's Honours List in 1946 and the End-of-War Honours List in 1946 – were rejected.[89]

---

[88] Citation dated 8 Nov 1945. Recommended for MBE for Brigadier N. Hamilton Fairley. NFX137899 Major Mabel Josephine Mackerras.

[89] Australian War Memorial, 'End of War' Awards submission by QMG and DGMS. Rejected Citation (MBE) for Major Mabel Josephine Mackerras NFX 137899 and for Major Winifred Iris MacKenzie VFX 81148, AWM 119–248/25; and Australian War Memorial, Recommendation for King's Birthday Honours 1945 (Office of the DQMG [Department of the Quarter Master General] & DGMS [Director General of Medical Services]). [Recommendations] Not Accepted [by the Chief of the General Staff]. Ref: AWM 119–266 (Part 2). Control No 147A. Barcode 5192561.

CHAPTER FIVE

The Vietnam conflict

If I ever have flashbacks it would be this: it's going into the ward and seeing how nurses will place a person in the bed so they're easiest to look after in intensive care ... to go into ICU and see a hale and hearty body put halfway down the bed because they don't have any legs I think is something I will never forget.1

Di Fairhead (née Skewes) – the first physiotherapist appointed
to the 1st Australian Field Hospital in Vietnam.

The Pacific War was to have a deep, enduring influence on Australia's military operations for the next two decades and beyond. It shifted the nation's focus from distant battlefields in Europe and the Middle East to the Asia-Pacific region and Australia's immediate neighbours in South-East Asia.

The fear of Japanese invasion had been real during World War II even though its likelihood remains a point of contention among military historians.[2] The bombing of Darwin on 19 February 1942 was the

---

[1] Extract originally sourced from interview transcript with Di Skewes available at: www.australiansatwarfilmarchive.gov.au.aawfa/transcripts/1190.aspx Verified in interview by Sharon Mascall-Dare with Di Skewes on 3 September 2013.

[2] The notion of a 'Battle for Australia' has galvanised debate concerning the likelihood of a Japanese invasion. While the RSL and veterans' groups have argued that Australia was forced to defend itself against a Japanese advance both at home (following air raids on northern coastal targets) and overseas (in New Guinea and elsewhere), Peter Stanley has argued that no such battle took place and is an attempt to make sense of Australia's devastating losses, primarily in 1942 and 1943. For an overview of the debate visit: http://www.pacificwar.org.au/battaust/

first of some 97 air raids on Australia's northern coastline, contributing to a climate of fear and insecurity. The attacks wreaked untold damage, both physical and psychological, although the extent of their destructive impact was largely unreported at the time. A new awareness of Australia's geography and potential vulnerability now prevailed: there was no guarantee that our Allies, on the other side of the world, could be relied upon in a crisis. There was a reorientation of Australia's strategic foreign policy, with consequences for the Australian Army.

The immediate concern, in the 1940s and 1950s, was security in the Asia-Pacific region and fear of communist expansion. Australian governments were determined to prevent neighbouring countries from falling, again, under the control of hostile governments and employed a policy of 'Forward Defence' to safeguard security in countries it perceived to be under threat.[3] This motivated Australia's involvement in Malaya, supporting Britain's counter-insurgency operations, and its involvement in Borneo during the Indonesian Confrontation. It also led to commitments to United Nations (UN) operations, including the Korean War, where personnel from the Australian Army, the Royal Australian Air Force and the Royal Australian Navy fought as part of a UN multinational force to defend South Korea from the Communist north. It also motivated Australia's involvement in Vietnam.

Australia and the war in Vietnam

Until the recent war in Afghanistan, the Vietnam War was the longest major conflict involving Australians, lasting for 10 years from 1962 to 1972. At first, Australia made a commitment of just 30 military advisers to the war effort; this grew to a battalion in 1965, and in 1966 a Task Force. By the end of the war more than 60,000 Australian men and women had served in Vietnam.

In the early years Australia's participation was not widely opposed. This

---

AustInvasion/Response_revisionists2.html Accessed 29 July 2013.

[3] For a comprehensive brief concerning Australia's 'Forward Defence' policy see: http://www.defence.gov.au/strategicbasis/pdf/1964.pdf Accessed 29 July 2013.

changed however, as National Servicemen were recruited and deployed to meet Australia's growing commitment. There was anger as the horrors of modern warfare were brought home through film and television. For the first time, war was watched in living rooms across the nation; it was no longer distant; it was real, it was here. As the public witnessed a war they thought was lost, opposition grew. In the early 1970s, more than 200,000 people marched in the streets of Australia's major cities in protest.[4]

Although recruitment to National Service targeted men and not women, a number of female volunteers found ways to contribute. They served as nurses in military and non-military capacities; they fulfilled civilian roles as entertainers, journalists, humanitarian workers or as consular and secretarial staff. Some had been part of the anti-war movement when they chose to volunteer.[5] As Siobhan McHugh discovered, when recording the oral histories of some 50 women for her study *Minefields and Miniskirts*, many experienced trauma:

> Alongside humorous and uplifting moments, several of the interviewees canvassed traumatic experiences. Civilian nurses described treating children burnt in napalm, or watching them die on the operating table. Military nurses spoke of seeing young men virtually castrated by landmines and of having to ring their girlfriends or family on their behalf.[6]

In addition to such horrors, medical staff contended with a lack of resources, an issue that prevailed in military as well as civilian contexts. In the *Official History of Australia's Involvement in Southeast Asian Conflicts 1948–1975*,[7] frequent comparisons are made between the limited medical resources provided by Australia and the support offered by the United States. As the historian Brendan O'Keefe has noted in his extensive study

---

[4] See http://vietnam-war.commemoration.gov.au/vietnam-war/index.php

[5] McHugh, Siobhan, '*Minefields and Miniskirts:* the perils and pleasures of adapting oral history for the stage,' *Oral History of Australia Journal*, no. 28, 2006.

[6] ibid, p. 22.

[7] O'Keefe, Brendan, 'Medicine at War: Medical Aspects of Australia's Involvement in Southeast Asian Conflicts 1950–1972', *The Official History of Australia's Involvement in Southeast Asian Conflicts 1948–1975*, Allen and Unwin, Australia, 1994.

of military medical operations during this period, Australia's decision not to commit more substantial resources had historical origins:

> Within the Australian Government and, at times, senior levels of the Australian military, there was a tendency to think of the armed forces as composed essentially of combat units and thus to neglect supporting arms and services such as the medical services. The attitude dated from World War I when the Australians had mounted an expeditionary force that, in comparison to other forces, had contained an abnormally high proportion of fighting units.[8]

O'Keefe also notes that, while generous, the medical support offered by the US military "reinforced a culture of dependence" within the Australian Government with repercussions for the services provided by Australia. The RAAMC had made significant advances in the treatment of malaria. Major Josephine Mackerras and her husband, Ian, had contributed key discoveries in this area, as part of Brigadier Fairley's unit, as had anti-malaria drug trials in the Atherton Tablelands.[9] This research played an integral role in the suppression and management of the disease, with ongoing repercussions for those Australians who served in South-East Asia. The disease had taken its toll on Australian service personnel during World War II and research programs had responded accordingly; still, the disease continued to takes its toll on Australian personnel serving in subsequent conflicts in the region.

It was disease, therefore, rather than injuries sustained on the battlefield that preoccupied Australia's medical teams initially. Total hospital admissions among Australian Forces reached around 13,000 for the duration of the war; of these 2,012 were for injuries sustained in battle and 2,650 were non battle casualties. The remainder were disease related.[10] Sexually Transmitted Disease (STD) was a significant problem, but most

---

[8] ibid, p. 273.

[9] See: chapter 3.

[10] Bentley, Philip and Dunstan, David, *The Path to Professionalism: Physiotherapy in Australia in the 1980s*, Australian Physiotherapy Association, 2006.

cases were treated as outpatients; between July 1966 and December 1969 there were 463 STD hospital admissions.[11] Attempts at keeping Australian servicemen out of the brothels that proliferated in the southern port town of Vung Tau were "conspicuously unsuccessful", despite the use of provost patrols. In October 1966, the rate of STD among Australian soldiers had peaked at a rate of 943 cases per 1,000 men per year (324 cases in 4,123 personnel). According to O'Keefe:

> There were various reasons for the failure of control. As the number of prostitutes and brothels in Vung Tau swelled, the job of policing the town moved steadily beyond the capabilities of the small provost strength.[12]

The problems were exacerbated by ingrained Vietnamese corruption in Vung Tau at all levels. Alcohol was a significant problem because of its innate ability to reduce inhibitions and increase bravado with the omission of wearing condoms.

As a consequence, the incidence of gonorrhea, non-specific urethritis (chlamydia) and chancroid was high, leading to collaborative efforts between the Australian Army and local Vietnamese authorities to introduce health screening for local prostitutes in order to reduce the risk of infection. The treatment and prevention of STD remained a medical problem from 1966 until 1972. Battle casualties continued to be a more pressing issue.

By 1969, Australia's military medical teams had handled a malaria epidemic (in October–November 1968); they had also established health screening for brothel workers in Vung Tau. Now accustomed to dealing with frequent outbreaks of dysentery and hepatitis, they were also

---

[11] Data provided by Peter Byrne in private correspondence to the authors, 3 December 2013. Colonel Byrne served with the RAAMC in Vietnam from July 1969 until August 1970. A Medical Officer (MO) with the rank of Major, in 1969 he was Deputy Assistant Director Medical Services, Australian Force Vietnam (DADMS AFV), and acted as MO Australian Army Training Team Vietnam – Special Forces (AATTV). In 1970 he was posted 1 Aust Fd Hosp for the last three months of his deployment from June to August 1970 and he held the position of Second-in-command (2IC), succeeding Major Shirley Coghlan, who is mentioned in this chapter.

[12] ibid, p. 103. Such practices were not limited to Vietnam, they were also documented in Cambodia. See chapter 6.

dealing with psychiatric cases as war took its toll on a generation of young men.[13] An increase in battlefield casualties brought a sudden increase in workload at the 1st Australian Field Hospital (1 Aust Fd Hosp[14]) located at Vung Tau. There was a a medical specialist staffing crisis with no prospect of immediate respite.

Raised on 1 April 1968 in Vung Tau, 1 Aust Fd Hosp expanded the services provided initially by the 2nd and 8th Field Ambulances. Established by the RAAMC in 1966, the 2nd Field Ambulance initially comprised half a stretcher-bearer unit and a 50-bed unit. It also had a forward company based at Nui Dat, the location of Australia's major military base in South Vietnam, some 20 miles inland.[15] From the outset the workload for Army medical teams had been variable at best; at worst, it became overwhelming:

> Battle casualties in Vietnam presented problems not previously encountered in armed conflict by Australian Army medical personnel. Helicopters delivered the wounded to 2nd Field Ambulance about 20 minutes after injury. Some of these soldiers were severely shocked, but alive, whereas in previous conflicts they would have died before reaching a major medical unit. Modern weapons were producing severe multiple contaminated wounds with much greater tissue damage than previously experienced. Surgery was frequently performed at the same time as resuscitation.[16]

By 1969 1 Aust Fd Hosp had expanded to 106 beds comprising a 50-bed surgical ward, a 50-bed medical ward and a six-bed Intensive Care Unit (ICU).[17] Between March and August that year, the number of

---

[13] ibid, pp. 103–4.

[14] Although the abbreviation '1AFH' is occasionally used, informally, to denote the 1st Australian Field Hospital, it is incorrect. 1 Aust Fd Hosp is the correct abbreviation.

[15] This data, previously recorded in the official history of 1AFH, has subsequently been removed from the official 1AFH Association website. It is, however, available via internet archives at: http://web.archive.org/web/20080908075319/http://callsignvampire.org/content/view/33/37/ Accessed 31 July 2013.

[16] Bruce, Greg K., 'Deployment of orthopaedic surgeons on ADF missions', *ADF Health*, 2001; 2(2): 80–84.

[17] Byrne, see footnote 11.

battle casualties doubled – the result of increased use of fragmentation weapons such as rocket-propelled grenades and mines. Closure of the 36th Evacuation Hospital, the major US medical facility in Vung Tau, in October that year put further pressure on the Australians. Australian Defence Force surgical teams now had to cope with all casualties, including severe injuries previously treated by the Americans.[18]

It was here, in July 1969, that a young woman from Adelaide was given the task of setting up a physiotherapy department from scratch, a challenge that she was not expecting. Aged just 25, Lieutenant Di Skewes (later Fairhead) had no prior experience of working in a warzone. During her one year tour-of-duty, she demonstrated the same ingenuity and resilience as other women who had served with the RAAMC before her.

Lieutenant Dianne Skewes

Di Skewes completed her three-year diploma in physiotherapy in Adelaide in 1965 and spent a year working at the Royal Adelaide Hospital. Her plan was to combine short-term locum work with travel: she arrived by ship in the United Kingdom (UK) and worked at the National Hospital for Nervous Diseases, Queen's Square, a renowned centre for neurological excellence and an ideal training ground for a newly qualified physiotherapist.

During her time in the UK, Di undertook locum work in Scotland as well as London. At one point she worked in the burns unit at Bangour General Hospital between Edinburgh and Glasgow, a position that proved to be particularly instructive. Di was confronted by horrific injuries that, with hindsight, prepared her for her work in Vietnam. "It was another situation that I went into and found that I could function," she recalls. "It was the ability to see these absolutely shocking injuries, but still be able to do what's expected of you without falling in a heap."[19]

In 1969, after 18 months away, Di felt it was time to come home. "I really had to force myself to come back to Australia," she recalled. "Eventually I thought, 'I'm getting too comfortable here … forgetting

---

[18] Bruce, G.

[19] Extract from interview by Sharon Mascall-Dare with Di Skewes on 4 October 2013.

Lieutenant Dianne Skewes in uniform before her deployment to Vietnam in 1969.
Image courtesy of Dianne Fairhead (née Skewes)

what life was like in Australia and I should probably go home and see what it's like, and then if I want to come back to the UK, I will." Her decision to return took her elsewhere, however. In July that same year she was posted to Vietnam.

During her time in the UK, Di had heard little about the war. She knew little about the controversy surrounding Australia's involvement; the British newspapers made scant reference to affairs 'Down Under'. Certainly, she was not aware that the war had become such a divisive issue. On her return to Australia she looked for work and saw an advert in the *South Australian Physiotherapy Newsletter*: "Physiotherapist required for 12 months' service in Vietnam urgently."[20]

Not realising that the job was with the Army, Di responded. She learnt the identity of her prospective employer when she was invited to undertake the Army's usual medical and 'psych' interviews. "I found out that it was in the Army," she said. "At that stage I was not strongly for or against

---

[20] See footnote 1.

The Vietnam conflict

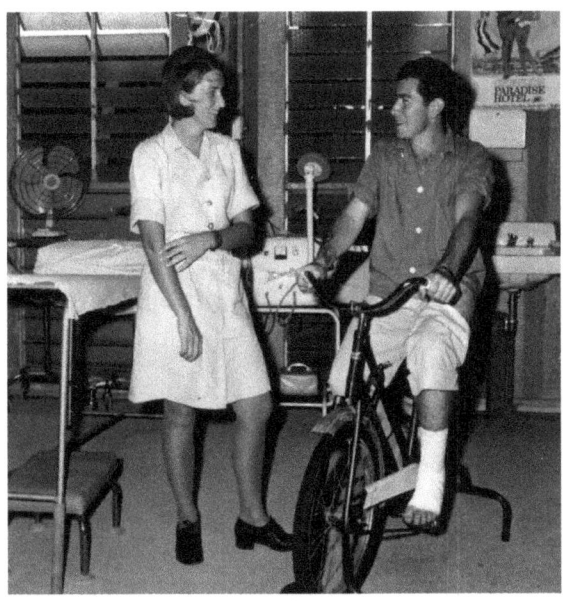

Lieutenant Dianne Skewes consults with a patient in the 'out-patient department' she established from scratch at 1 Aust Fd Hosp. She requested an exercise bike, some free weights and a traction machine. She also requested a fan. *Image courtesy of Dianne Fairhead (née Skewes).*

Australian soldiers being in Vietnam, but figured if they needed a physio I was happy and able to be in it."[21] After she passed initial screening in Adelaide, she flew to Melbourne for an interview. She was selected that day and enlisted the same week.

Before her departure for Saigon, Di spent six weeks at the 2nd Military Hospital at Ingleburn in New South Wales. It was an opportunity to learn military protocol and talk to medical officers and nurses about the realities ahead. She was also waiting for her uniform that was, effectively, tailor made from whatever could be found. "Do I have to have a uniform?" she asked, "can't I just go?"[22] Her day uniform comprised a long-sleeved light-brown cotton dress, worn with the sleeves rolled up, together with

---

[21] Information originally sourced from Wilson, Honor, C., *Physiotherapists in war: the story of South Australian physiotherapists during World Wars I and II, Japan–Korea and Vietnam*, Hazelwood Park, SA, 1995, p. 151. Updated in an interview by Sharon Mascall-Dare with Di Skewes on 4 October 2013.

[22] Extract from interview by Sharon Mascall-Dare with Di Skewes on 4 October 2013.

stockings and black lace-up shoes. Since the requisite brown felt hat was not readily available, the Army found one in war storage, dating from 1944. It was duly issued to Di, with the customary embellishments of the RAAMC.

Di was also provided with 'greens' for Vietnam: green trousers, shirts, 'GP' (general purpose) boots and a helmet. Her luggage allowance included a trunk, a small leather suitcase and green canvas bag – an issue almost unchanged from World War II.[23] She flew to Saigon by Qantas charter jet and then by RAAF Caribou to Vung Tau. She arrived at 1 Aust Fd Hosp in early July 1969 and remembers the base as, "Very austere, military … a business-like place. It was not like anywhere else I'd ever worked. At least at Bangour [in Scotland] there were a few trees, there was colour".[24]

Di was the only physiotherapist working at the hospital. Her first task was not one she had expected. She was to establish a functioning 'out-patient department' from scratch, at the end of a Nissen hut with a cement floor. There was no airconditioning and equipment had to be ordered from the Americans. "The CO [Commanding Officer] said: 'What do you want, give us a list and we'll get it'", she recalls. "I was a bit shocked that I had to order all this equipment. It then appeared magically from the Americans".[25] Di requested an exercise bike, some free weights and a traction machine. Importantly, given the heat, she also requested a fan.

Prior to her deployment, Di had met a number of surgeons who had recently returned from 1 Aust Fd Hosp. They were concerned about the rise in chest complications among soldiers in intensive care and believed that physiotherapy could help. This was also raised during her initial Army interviews. It was a capability that had been missing from the intensive-care team at 1 Aust Fd Hosp. Clearing mucus secretions from the men's chests played a crucial role in their recovery, leaving them far less likely to contract infections such as pneumonia, which could be deadly.

Lieutenant Skewes's primary responsibility, therefore, was to provide chest physiotherapy for patients in intensive-care. It was hard work for

---

[23] Wilson, p. 151.

[24] See footnote 1.

[25] Extract from interview by Sharon Mascall-Dare with Di Skewes on 4 October 2013.

both patient and therapist, and required mutual support between Di and other members of the intensive-care team. Di's work required persistence as well as strength:

> Physio in intensive care is not nice: it is making you uncomfortable; that's what we're supposed to do. On one occasion, I can remember there was one guy in there who had a really bad chest and I'd been giving him a really hard time because he was pretty well consolidated, which meant he had pneumonia on one side of his lung. So that meant I had to do a lot of work with him to try and get air back into that area and move him and get him to cough when everything else hurt because [he] had multiple fragment wounds in [his] abdomen. If you think of anyone who's got a fractured rib or a little incision, coughing is the worst thing they want to do … I'd been working with him and he was slowly improving and another soldier was admitted who needed chest physio. The fellow who I thought I'd made my absolute worst enemy said, "Work with her, it's good, you'll feel better afterwards, it's dreadful at the time but you'll feel better afterwards, it's worth it". And I thought, "Well that's good: that's the sort of feedback you probably need".[26]

Resourcefulness was also important. While Di's expertise was badly needed at the hospital, there was no-one for her to talk to about establishing a new physiotherapy department at 1 Aust Fd Hosp. "Now, when I think about it, I cannot believe that I did it … going into the absolute unknown," she said. "I'd had quite a lot of experience in different places, which does help going in a new situation, not knowing anyone and just being able to pick up and work … I'd obviously managed to do that [in Europe] so I suppose they thought that I might be able to do it there."[27]

Almost all of her patients were Australians and New Zealanders injured by gunshot wounds and mines. She also saw a large number of minor sporting injuries, sustained away from the fighting, back on the base. More

---

[26] See footnote 1.
[27] ibid.

serious cases were stabilised before evacuation back to Australia, where longer-term rehabilitation took place. To her disappointment, Di was not involved in this phase of recovery: "When they're stabilised, that's when the rehabilitation can commence. I was frustrated that I could not be involved in the good part of their rehabilitation".[28]

At times, Di also treated American soldiers, and occasionally, members of the Vietcong who had been captured as prisoners of war (PWs). To her disappointment she did not have the opportunity to treat civilians, as this was not covered by the hospital charter.[29] "I do recall, especially after going to a civilian hospital, the varied work that they were able to do. We could not since we were an Army hospital," she said.[30]

Di summarised her work as follows:

- Chest care in the Intensive Care Unit with soldiers injured in mine explosions.
- Minor gunshot wounds or fragment wounds requiring short-term rehabilitation.
- Sprained ankles, twisted knees, back pains after 'sporting injuries' from games at the base.[31]

Her working hours began at seven in the morning and sometimes continued until seven in the evening on busy days. She worked when she was needed: Monday to Saturday as a matter of routine and also Sundays when busy. On her days off she was sometimes allowed to take a trip by helicopter to Saigon or along the coast, usually with one of the American pilots. The Peter Badcoe Club for Allied servicemen was not far from the hospital and housed sporting facilities, a swimming pool, tennis courts and small yachts for sailing the South China Sea.

Female officers were a rare commodity and Lieutenant Skewes was often invited out with other off-duty women from the base. She was friends with a number of female nurses and 'Red Cross girls'. When the battalions

---

[28] Extract from interview by Sharon Mascall-Dare with Di Skewes on 4 October 2013.
[29] Wilson, p. 152.
[30] Extract from interview by Sharon Mascall-Dare with Di Skewes on 4 October 2013.
[31] Wilson, p. 152.

had rest periods at Nui Dat or Vung Tau there were parties, and Di was often invited, hitching a lift in a helicopter or RAAF Caribou. There were also opportunities to visit local markets and American clubs or French restaurants famed for their exorbitant prices. Outings were allowed as long as there were no 'Red Alerts' and Di returned by the 10 o'clock evening curfew. There were also rules to protect her safety:

> The security for the females was quite strict. We were allowed out into Vung Tau, but we had to go in 1ALSG [Australian Logistics Support Group] vehicles. We weren't allowed to just go out of the camp area and walk, that was a no-no.[32]

As a physiotherapist in the RAAMC, Lieutenant Skewes was answerable directly to her Commanding Officer (CO). This was similar to World War II, when the small number of physiotherapists only came under the matron's control for disciplinary matters. At times, this caused tension, as Di observed: "I was always treated as a valuable member of a very efficient team by the doctors and nurses. I personally had a good relationship with matron, but I knew it irked her that as a Medical Corps Officer I was answerable to the CO and not her as were all the other female nurses."[33] Fortunately for Di, her relationship with her CO was also good. He understood that she was somewhat isolated – in a team of one – as the only Australian physiotherapist working at the hospital. On one occasion he overruled the matron and granted permission for Di to attend the US Physical Therapist Conference weekend at Nha Trang, further north. It was a valuable opportunity to network with other physiotherapists and hear from guest speakers presenting on the role of physical therapy in the treatment of leprosy and recovery from burns.

Getting to the conference was a feat in itself. She flew north, along the coast, on the well-established 'Wallaby Run', a routine RAAF Caribou flight that had to negotiate difficult landings on local airstrips – the crew was never certain whether an airstrip was friendly or had fallen to the Vietcong. Due to rocket attacks on the town of Nha Trang itself, conference

---

[32] See footnote 1.
[33] Wilson, p. 152.

delegates were confined to the grounds of the US base and the 8th Field Hospital. Di felt fortunate to have seen so much of the country on her way there and back, even though the journey was punctuated by 'hairy' take-offs and landings.[34]

Although Di did not feel in danger, she was constantly reminded of the horrors of conflict. She treated amputees who had lost their limbs in mine explosions and was deeply disturbed by images of men placed halfway down the bed – a sign that their legs were no longer there. The sound of the aircraft and helicopters that filled the skies during the war stayed with her for years:

> It just gets in your bones, you just feel the vibrations, you just know. The spookiest thing I have ever experienced in my life was when they flew over at the dedication of the Vietnam War Memorial [in 1992] … there was this huge amount of noise on an oval when everyone was gathering and these Iroquois flew over, well you could have heard a pin drop … everyone is affected the same in some incredibly strong way by the helicopter, by the Iroquois. Everyone who served in Vietnam is affected by that sound.[35]

Towards the end of her time in Vietnam, Di became increasingly aware of the controversy surrounding the war, as thousands of Australians took to the streets in protest back home. She heard patients talk about letters they were receiving from home – not abusive, but questioning their involvement. "What in the hell are we doing here?" some of the men asked, "Why are we here? Why am I getting shot up?" As Di recalls, "It was certainly very, very difficult for them".[36]

Around this time Di met her future husband, Fred Fairhead, at a farewell party for 9 Battalion of the Royal Australian Regiment at Nui Dat. He had spent his career as an Army Officer and was Intelligence Officer

---

[34] ibid, p. 153.

[35] See footnote 1.

[36] ibid.

of the 6ᵗʰ Battalion.³⁷ He left Vietnam before the end of Di's deployment and he drove from Perth to Sydney to meet her when she returned. On the way, he made a brief stop in Adelaide to meet her parents and sister.

Di left the Army when she came home, married Fred and continued to pursue her career as a physiotherapist while bringing up a young family. Like many others who served in Vietnam, Di was reluctant to discuss her experiences at first, aware of the criticism surrounding Australia's involvement.

> That very special year full of amazing experiences was locked away with the letters and photos in a cupboard and the memories way back in my mind, because there was really no-one I could share them with as it had been such a unique experience.³⁸

Like others who served with her, she remembers feeling estranged from Australian culture on her return. "When I came back, people say 'How was it?'. They wanted a sentence, but there's no way you can explain it to someone who hasn't been there."³⁹

She was also shocked by a way of life she felt to be unnecessarily wasteful and materialistic. In Vietnam, she had become accustomed to making do with what she had; she had become resourceful. She was amazed at how much Australians ate and that they wanted for possessions they did not need.⁴⁰

In 1992, at the dedication of the Australian Vietnam Forces National Memorial in Canberra, Di allowed herself "the luxury of remembering",⁴¹ recalling a time that had polarised a nation and remained locked away in the minds of the men and women who served alongside her.

Now in her 70s, Di's memories have been recognised as a valuable resource and have been recorded as an oral history. Her experiences have been stored for posterity by Department of Veterans' Affairs as part of the

---

[37] Verified by Sharon Mascall-Dare with Di Skewes and Fred Fairhead on 4 October 2013.
[38] Wilson, p. 155.
[39] Extract from interview by Sharon Mascall-Dare with Di Skewes on 4 October 2013.
[40] Wilson, p. 155, see also footnote 1.
[41] Ibid.

Australians at War Film Archive. In recognition of her service career, Di has also been appointed to the board of Legacy in South Australia – the first woman to be given such a role.

"Legacy was once a purely male domain, but today there are many outstanding women as well as men volunteering their time to help families of deceased veterans," she told Adelaide's *Advertiser* newspaper on her appointment in 2011. "Whatever barriers to service there might have been have now come down."[42]

Paving the way

When Di left Vietnam in July 1970, she was replaced by Lieutenant Susan Woolley, a physiotherapist from Perth. The 'handover' was a brief chat between the two women in a transit lounge of Saigon Airport. Susan also flew to Vung Tau on an RAAF Caribou as Di had, one year earlier. Susan landed in a downpour in the midst of a monsoon and was met on the tarmac by the hospital matron from 1 Aust Fd Hosp who was "armed with an umbrella".[43]

Susan spoke highly of her time in Vietnam. By the time she arrived the physiotherapy department at 1 Aust Fd Hosp was well-established and she inherited a fully equipped unit that was part of a "well-organised and focused team". The hospital now housed 156 staff, including two surgeons, two anesthetists, a physician, pathologist, psychiatrist, three general duties medical officers, 11 RAANC nurses and two sisters from the Royal New Zealand Nursing Corps. Lieutenant Woolley was the hospital's only physiotherapist. Women were still in the minority. There were between 15 and 18 female members of staff. They were usually treated with "great courtesy".[44]

For Susan, the worst part was witnessing the "severe and disfiguring

---

[42] Kelton, Sam, 'Legacy of a female leader', *The Advertiser*, 3 December 2011, available at: www.adelaidenow.com.au/legacy-of-a-female-leader/story-e6frea6u-1226212744448 Accessed 15 June 2013.

[43] Bentley and Dunstan, p. 178.

[44] O'Keefe, and Woolley, Susan, Personal Communication, 4/10/1999, cited in Bentley and Dunstan, p. 179.

injuries sustained by the troops ... and knowing that the prevailing climate 'back home' was not sympathetic to Vietnam veterans".[45] On her return to Australia she continued to work with war veterans and found her experiences in Vietnam helpful: it gave her to ability to empathise with her patients.[46]

After 12 months' service, Susan was replaced by Captain Shirley Rae from New South Wales – the last physiotherapist to serve in Vietnam. The Australian Government began scaling down its presence in 1970 and almost all Australian troops had been withdrawn by the beginning of 1972.

OTHER MEDICAL WOMEN IN VIETNAM

After two years' part-time service with the Citizen Military Forces in Brisbane, Major Shirley Coghlan joined the Australian Army Medical Team in Vietnam as a specialist physician.

Major Coghlan had worked at the Repatriation General Hospital Greenslopes and the Princess Alexandra Hospital in Brisbane before she arrived in Vietnam in December 1969. She became the first female medical officer to serve officially with the Australian Army in a combat zone and was second in command at 1 Aust Fd Hosp from 13 February 1970 until 3 June 1970.[47]

The "much respected medical officer"[48] has been described as, arguably "the most capable specialist on the staff of the Field Hospital"[49] where she was in charge of post-operative intensive care:

> ... she produced some remarkable results. She was a calm modest woman with a wistful expression and a delightfully dry sense of humour ...

---

[45] Woolley, Susan, Personal Communication, 4/10/1999, cited in Bentley and Dunstan, pp. 178–9.

[46] ibid, p. 179.

[47] Williams, Lesley, M., *No Better Profession: Medical Women in Queensland, 1891–1999*, p. 161; dates and facts verified by Peter Byrne in private correspondence to the authors.

[48] Byrne, Peter, private correspondence, 3 December 2013.

[49] Spragg, Griffith, *When Good Men Do Nothing – The Memoirs of Griffith Spragg*, Australian Military History Publications, 2003, p. 93.

> Yet for a long time she was not well accepted or fully respected in some quarters of the unit; I can only assume because she was a woman in a man's world. Eventually her value was recognised, and the administration could not do enough for her.[50]

Two female anaesthetists and a female pathologist also served at 1 Aust Fd Hosp, either as contemporaries of Major Coghlan or arriving shortly after her departure. Anaesthetist Dr Rosemary Coffey, from Sydney, served for several weeks in late 1969 during the hospital's medical 'manning' crisis on loan from a civilian medical aid team. Another female anaesthetist, Dr Shirley Lee, enlisted in the RAAMC on 27 August 1970 and also worked at 1 Aust Fd Hosp. She left for Saigon on 3 September 1970, returning in December. Pathologist Major Barbara Grahame also served there in 1970.[51]

A number of civilian medical teams also also worked in Vietnam, continuing a tradition of military service – not necessarily in uniform – that can be traced back to the Scottish Women's Hospitals of the Great War.[52] They endured conditions in many ways more difficult than their uniformed counterparts. The first Australian civilian surgical team left for Vietnam on 3 October 1964 following a request from the Department of External Affairs. It asked the Royal Melbourne Hospital to supply a self-sufficient surgical unit comprising two surgeons, an anesthetist, a physician, two theatre sisters, a ward sister and a radiographer. The team was based at Long Xuyen, some 200 kilometres west of Vung Tau in southern Vietnam on the Mekong Delta. The provincial hospital in the town hosted Australian surgical teams until their withdrawal in 1970.[53]

Gary McKay and Elizabeth Stewart provide a comprehensive review of the contribution made by Australian civilian medical teams during the Vietnam War in their book *With Healing Hands*:

> Dealing with minor burns, broken limbs and mild illnesses, delivering babies, and treating patients with plague, tetanus,

---

[50] ibid.
[51] Details provided by Peter Byrne in correspondence to the authors; see also Williams, p. 161.
[52] See chapter 2.
[53] McKay, Gary and Stewart, Elizabeth, *With Healing Hands*, Allen & Unwin, 2009, p. 23.

malaria, advanced cancer, horrific war wounds and severe facial deformities, the teams worked through floods, enervating heat, and enemy attack to bring medical relief to an exhausted and traumatised Vietnamese population. The nearly 500 Australian men and women who signed up to work on surgical teams during the eight years they operated in Vietnam did so for a variety of reasons: adventure, excitement, professional challenge, empathy for a people at war. All found themselves at times confronted, scared, angry, exhausted and challenged, but few, if any, regretted their decision to go to Vietnam.[54]

The role of first team to arrive – from the Royal Melbourne Hospital – was similar to those that followed. They were not there to take over the hospital in Long Xuyen, but to improve the level of surgical support available. The team's activities were limited to the surgical units and pre- and post-surgical care, including radiography and anaesthesia.[55]

Conditions in the hospital were "primitive". Patients often slept two or three to a bed and sewage disposal was scanty. Forced to adapt to the conditions, the Australian team tried to improve conditions, leading by example.

The first radiographer to arrive with the Royal Melbourne team in October 1964 was Noelle Laidlaw (née Courtney). She found the hospital's radiography machine to be 'pretty ancient', with no X-ray films or developing fixer available. Frequent power cuts made her job even more difficult. She ordered supplies from the Royal Melbourne Hospital and local Vietnamese set to work on repairing the darkroom after she complained that there was no exhaust fan or running water.[56]

Noelle performed other roles at the hospital when the X-ray machine broke down; she became multi-skilled, playing an important role in the establishment of a blood bank, for example. Local Vietnamese people were reluctant to give blood, even if a family member's life was at stake; they believed that donating blood would leave them weak and their bodies

---

[54] ibid, p. xv.
[55] ibid, p. 28.
[56] ibid, p. 30.

would not recover. As McKay and Stewart describe:

> The Australians realised they had a difficult task on their hands. Only a few weeks after her arrival, Noelle Laidlaw (who helped Dr Tim Matthews oversee the blood-bank project) found herself donating blood to a man who had been brought in with extensive gunshot wounds to the abdomen. A mixture of her own, American and Vietnamese blood was used during surgery to save the man, who would otherwise have died. The fact that she showed no ill effects and that the patient recovered helped the blood-bank idea take hold.[57]

Noelle left Vietnam in January 1965. The next female radiographer to work at Long Xuyen was Barbara Maughan (later Sutherland), who arrived in October 1966 and left in April 1967. She was part of a team from Prince Henry's Hospital in Melbourne.

In Vung Tau, the first civilian surgical team arrived in November 1966. Its personnel came from the Prince Henry and Prince of Wales hospitals in Sydney. Members of the team were chosen by its leader, the head of surgery at the Prince Henry, Professor Doug Tracy. They included a female anesthetist, Judith Ross, who worked in Vung Tau until February 1967.[58]

THE LEGACY OF VIETNAM

In all, 500 Australian troops died and 3,129 were wounded in the Vietnam War.[59] For the US, the death toll was much higher, reaching 58,220.[60] The total number of deaths among Vietnamese (civilian and military, including both North and South Vietnamese) was higher still. Demographic analyses have estimated the death toll to be between 791,000 and 1,141,000.[61]

---

[57] ibid, p. 32.

[58] ibid, p. 125.

[59] Sourced from the Australian War Memorial, available at: http://www.awm.gov.au/encyclopedia/vietnam/statistics/ Accessed 2 August 2013.

[60] See http://www.archives.gov/research/military/vietnam-war/casualty-statistics.html Accessed 16 June 2014.

[61] Hirschman, Charles, et al., "Vietnamese Casualties During the American War: A New

The Vietnam conflict

The collapse of the South Vietnamese Government, following the withdrawal of American troops in 1975, triggered a humanitarian crisis that continued to have repercussions for decades to come. In the days before the fall of Saigon, 140,000 Vietnamese who were closely associated with the South Vietnamese Government were evacuated to the United States; a further 12,000 refugees fled to Thailand, Hong Kong, Singapore and the Philippines before the year was out. As the numbers of refugees grew from tens of thousands to hundreds of thousands, the Asia Pacific region struggled to cope with the scale of the crisis.

The next four years saw a mass exodus of 'boat people' from Vietnam, Cambodia and Laos. According to the United Nations High Commissioner for Refugees, around 1.3 million refugees were resettled between 1975 and 1997; many more died at sea. Australia took a lead role in the resettlement of refugees, alongside the United States, Canada and France.[62] Entire suburbs of Australia's largest cities were transformed.

The legacy for individuals – as well as communities – remains difficult to quantify. While some veterans of the war experienced exclusion and hostility on their return, others did not; while some refugees built new lives in Australia, others – along with veterans – were haunted by trauma. For some, memory is embedded in the sound of Iroquois helicopters; for many it is remembered as they witness another generation of refugees fleeing a new generation of conflict.

For another young woman, only in her teenage years, it shaped a series of events that would see her, as a refugee of Vietnam, put on the uniform of the Australian Army and serve with the RAAMC.[63]

Estimate," *Population and Development Review*, Vol. 21 [4], December 1995, pp. 783–812.

[62] UNHCR, 'Flight from Indochina', *The State of the World's Refugees 2000: Fifty years of Humanitarian Action*, 1 January 2000, pp. 84–6.

[63] See chapter 7: Captain Tran CSM

CHAPTER SIX

Other people's wars: peacekeeping near and far

> *Today, we have more than 110,000 men and women deployed in conflict zones around the world. They come from nearly 120 countries ... They bring different cultures and experiences to the job, but they are united in their determination to foster peace. Some are in uniform, but many are civilians and their activities go far beyond monitoring. They train police, disarm ex-combatants, support elections and help build State institutions. They build bridges, repair schools, assist flood victims and protect women from sexual violence. They uphold human rights and promote gender equity. Thanks to their efforts, life saving humanitarian assistance can be delivered and economic development can begin.*
>
> Ban Ki-Moon, eighth and current Secretary-General of the United Nations.[1]

At the end of the Vietnam War, Australia fell into what was colloquially called "The Long Peace". Influenced by fears of communism and a Cold War enemy, it was a time when the Australian Army focused on preparations for a conventional war.

Over the ensuing decades, the Army contracted and, other than a brief engagement in the First Gulf War,[2] it became an era of peacekeeping.

---

[1] New York, 29 May 2008 – Secretary-General's message on International Day of UN Peacekeepers. Available at: http://www.un.org/sg/statements/?nid=3191 Accessed 13 June 2014.

[2] See: chapter 7.

Other people's wars: peacekeeping near and far

Australians had been among the first United Nations peacekeepers anywhere in the world, having first deployed to Indonesia in 1947.[3] From 1990 onwards Australia participated in what became the busiest decade in the history of multinational peacekeeping, with a commitment to peace and security, not only in our own region, but elsewhere around the globe.

Australia took a leading role in United Nations operations in Cambodia and Somalia. Australian peacekeepers have served in the remote desert lands of the Western Sahara and the hillsides of Rwanda; they have been deployed in the mountains of Kashmir and on the Iran–Iraq border.

Many of these missions have been arduous and dangerous. In almost all, medical women of the Australian Army have served among them.

From 1997, Australians served in regional peace-monitoring operations in Bougainville and the Solomon Islands. These efforts were dwarfed however, in 1999 when, in response to East Timor's independence from Indonesia, Australia led an international peace-enforcement operation.

In 1993, women made up 1% of uniformed personnel in United Nations missions.[4] For Australian women however, this percentage was significantly higher.[5] This was an era in which opportunities for women in the Australian Army changed markedly. Female medics were deployed alongside their male counterparts as equals. Female doctors deployed as Regimental Medical Officers and specialists. Women commanded medical teams and hospital facilities overseas.

These women have demonstrated the same resilience and willingness to use their skills as the women in World War I. They too have had to adapt, demonstrate flexibility and improvise in often difficult, complex and dangerous circumstances. Unlike their forebears however, these women deployed wearing the uniforms of the Australian Army and the embellishments of the RAAMC.

---

[3] See: https://www.awm.gov.au/atwar/peacekeeping.asp Accessed 13 June 2014.

[4] Consolidated Statistical Information on Female Military and Police Personnel in UN Peacekeeping Operations from 2005–2010. Available at: http://www.un.org/en/peacekeeping/documents/gender_scres1325_chart.pdf Accessed 13 June 2014.

[5] For a period in 1993, Australia had over 2,000 peacekeepers in the field, with large contingents in Cambodia and Somalia. In Cambodia alone in 1993, of a 600 strong contingent there were eight female officers and over 30 female soldiers.

Cambodia

On a warm afternoon in late 1992, Colonel Wayne Ramsey sat in the gracious sandstone buildings of Army's Land Headquarters at Paddington Barracks in Sydney.

With increasing political rhetoric about Australia's ongoing involvement in the largest UN peacekeeping mission to that date – and the risks it posed to over 600 Australian soldiers that were serving as part of the United Nations Transitional Authority in Cambodia (UNTAC) – Colonel Ramsey had a problem.

Politicians were concerned about a Vietnam-style quagmire; the public was concerned about the possibility of increased violence.[6] The spectre of Australian casualties was growing – Cambodia's minefields were a deadly, pervasive enemy. Colonel Wayne Ramsey, Director General of Army Health Services was however, concerned about something else.

Weekly medical SITREPS[7] from Cambodia revealed rising rates of sexually transmitted disease, not only across the UN, but also among the Australian and New Zealand contingents. Prostitution was on the increase, encouraged by the presence of so many international troops in the country. The international community was outraged. While most STDs were treatable, HIV–AIDS was not, and the prospect of an Australian soldier contracting HIV–AIDS was becoming a looming reality.

Australia's contribution to peacekeeping operations in Cambodia was significant. Between 1975 and 1978 the Khmer Rouge, under the leadership of Pol Pot, had undertaken one of the most brutal 'social reconstruction' programs ever attempted. Cambodia was cut off from the outside world; currency and postal services ceased. Hospitals, offices and schools were closed. Towns were emptied to form slave-labour teams. An estimated 1.7 million people were killed by a genocidal campaign that targeted government employees, the educated and anyone who resisted. Thousands

---

[6] The London-based Australian journalist John Pilger was the first reporter to access Phnom Penh after the Vietnamese removed Pol Pot in 1979. His confronting documentaries were instrumental in shaping Australian public opinion: *Year Zero: The Silent Death of Cambodia* (1979) and *Cambodia Year Ten* (1989).

[7] SITREP is an Army acronym for situation report – a summary of the current situation as experienced by a unit or formation, or during an operation, mission or activity.

died from disease and starvation, forced to undertake backbreaking work in the fields with little more than thin, watery rice porridge for sustenance. Vietnam's 'liberation' of Cambodia in 1979 failed to restore peace. Reprisal attacks, show trials, famine and ongoing factional fighting continued to take its toll on a country devoid of infrastructure.

Australia, under the stewardship of Foreign Minister Gareth Evans, played a key role in establishing peace talks in Paris in 1991. It also obtained international agreement for a UN mission to oversee disarmament and pave the way to a legitimate electoral process in Cambodia. In October 1991, the Australian Parliament agreed to commit 65 personnel to the UN Advance Mission in Cambodia (UNAMIC), including signallers, military observers and support staff.

Six months later, the UN Transitional Authority in Cambodia (UNTAC) was formed. The Force Commander was an Australian, Lieutenant General John Sanderson. Australia's military commitment to UNTAC was two 12-month rotations of personnel deployed to the Force Communications Unit (FCU). Overall, the international mission comprised 12 infantry battalions, support units, military observers and civilian police – a total of 22,000 personnel from 32 different countries.

ANZAC Ward

The Australian and New Zealand contingent was supported by a small medical facility based in Phnom Penh with the main body of the FCU.

The medical facility comprised a ward next to a Regimental Aid Post or RAP – the stalwart of any battalion or regimental health centre. There were half-a-dozen medics on staff, alongside two lab technicians, a pharmacist and a nursing officer. The first RMO to command the team was Captain Peter Roessler, who left after a 12-month deployment. He was replaced by a female doctor in March 1993 – Captain Susan Evans (later Neuhaus), co-author of this book.[8]

Captain Evans was 27 years old. She had joined the Army as an undergraduate medical student in her fifth year at the University of

---

[8] Susan Neuhaus's experiences in Cambodia are included in the epilogue.

Adelaide. After an intern year at the Royal Adelaide Hospital and an 18-month surgical residency in the United Kingdom, she had been posted to the 1st Recruit Training Battalion, Kapooka; the formative training ground for soldiers of the Australian Army.

Quelling any concerns about a female medical officer, Colonel Ramsey replaced Captain Annie Blundell with a male nurse, Captain Louis (Lew) McLeod. The irony of this gender reversal was not lost on the regiment; many a soldier would ask to see the 'male doctor'.

Daily routine at the RAP involved ward rounds and morning sick parades, where medics would see soldiers requiring minor treatments or refer them on to the doctor. Afternoons were reserved for education sessions, visits to orphanages and medical boards – compulsory medical clearances for every soldier entering or leaving the country to reassess their vaccination status, and screen for malaria or HIV. [9]

'ANZAC Ward' had six stretcher beds elevated on frames – iron loops on the sides served as anchor points to secure mosquito netting. The RAP had a single resuscitation bay, the Regimental Medical Officer or RMO's office, a pharmacy and a small, transportable, airconditioned laboratory. The heat was oppressive. From December to January the humidity levels were relatively low, but the monsoonal buildup from April through to June made for testing times as temperatures soared above 40 degrees and generators struggled to cope. By far the largest object in the RAP was a 400-litre glass-fronted refrigerator. It was brimming with anti-venom and carried useful 'photo ID' to identify any offending snake.[10]

Malaria and Tropical Disease

The threat of malaria and other tropical diseases was constant. Drawing on the expertise of the Australian Army Malaria Research Unit, the FCU medical facility was also tasked as the United Nations malarial 'reference laboratory', and cross-checked every smear from UN and local hospitals

---

[9] This was a considerable workload, particularly at times when detachments rotated their staff back. As the UN withdrawal commenced, a second doctor was brought from Australia to assist.

[10] Despite the reputation of the King Cobra it was actually the Haluman snake that was most to be feared: small but deadly, it is light green and emerges after rainstorms.

that was suspected to be positive for malaria. The contribution of this small team to international efforts by the UN and World Health Organisation to control malaria in the country was significant.

More problematic than malaria was dengue, a viral disease transmitted by a seasonal, urban and daytime biting mosquito. A vague febrile illness would be followed by muscle aches and pains, with telltale small red spots appearing about four days later, usually on the trunk. The virus is usually self-limiting, but on rare occasions the complications of dengue hemorrhagic fever can be fatal.

In some parts of the country bilharzia was still prevalent and soldiers had to be reminded of the risks of swimming in infected waters. The disease is caused by minute Schistosomal parasitic worms that thrive in freshwater rivers and lakes, where they enter the body through the skin to attach to the bladder and intestines.

Rabies was also an issue, primarily when soldiers neglected advice not to get too close to monkeys or 'adopted' one as a detachment mascot. A number of soldiers experienced the painful reality of rabies anti-toxin, injected directly into a bite wound to block the virus from spreading through the nervous system.

The Australian medical team did its best with limited resources. Lacking a full range of diagnostic testing facilities, much of the medical treatment was empirical; bed rest, fluids, antibiotics and constant observation for a rising fever or rash.[11] Despite the RAP's collection of 'Tropical Disease' textbooks, not much of medical-school training proved to be of use 'in country'. Fortunately, the majority of the workload was more straightforward. Sporting injuries, skin rashes, diarrhoea and traffic accidents were common in Phnom Penh where fleets of white UN vehicles added to the already chaotic traffic.

Pressure on the team grew as Cambodia prepared for elections in 1993. Casualties were expected from outbreaks of violence; Khmer hospitals were poorly resourced and local blood supplies were precariously low. In response, the FCU medical team transformed ANZAC Ward into a community blood bank. Australian soldiers donated blood and Khmer

---

[11] Often a cocktail of ciprofloxacin and quinine.

technicians, using a technique unchanged since World War I, performed cross-matching on porcelain tiles. More than 150 units of blood were obtained and labelled in this way.[12]

Landmines

Landmines were a constant threat throughout Cambodia. Left over from both the Vietnamese occupation and the reign of Pol Pot, they littered the country, buried in ricefields and by roadsides.

A favorite tactic of the Khmer Rouge was to lay mines across roads and paddy fields, deliberately targeting civilian communities and disrupting village life and their reliance on agriculture. Areas that appeared safe in the dry season became unsafe following the seasonal rains as the earth softened and the mines moved.

Cambodia still has one of the highest amputee rates per capita in the world (1:275), with children starkly overrepresented. Despite warnings, children would play with the objects they found floating in the paddy fields, not knowing the danger. Children are particularly vulnerable to the effects of mines due to their small stature; they are more likely than an adult to lose a hand or arm or their vision. In the 1990s Cambodia was also home to the world's largest single minefield – K5 – extending from the Gulf of Thailand to the Laotian border.

Just before Easter 1993, a group of Australians had a lucky escape after rolling their Landrover into a minefield south of Phnom Penh. Only one soldier sustained a major injury – a compound fracture of the femur. He spent several days in ANZAC Ward with his leg in makeshift traction – built by FCU engineers using gym weights and bags of saline. The RAAF then sent a C-130 Hercules to transport him back to Australia for surgery. The padre, who was also travelling in the vehicle, sustained a broken nose.

---

[12] Australian soldiers donated blood, but were not recipients. No Australian received a transfusion during the UNTAC deployment, but emergency stocks of O negative blood, sourced from the Australian Red Cross were held in reserve at the RAP.

Sexually Transmitted Diseases

As the SITREPs showed, one threat did not come from landmines or tropical diseases; it came from sexually transmitted diseases (STDs). As on previous operations, a minority of deployed personnel had gonorrhoea, herpes, syphilis or non-specific urethritis (NSU).[13] These diseases however, were treatable: HIV–AIDS was not.

In the 1990s, HIV–AIDS was a feared disease, even among members of the scientific and medical community. Television images of the Grim Reaper with his bowling ball – representing the indiscriminate threat of HIV–AIDS – gripped public consciousness. At the time there was no effective treatment; a diagnosis was a death sentence. Concerns about the virus spreading through the Australian community bordered on hysterical. Although the RAP could test for HIV–AIDS, the screening kits were notoriously oversensitive and required additional formal testing in Australia. These results took up to four weeks to return; an agonising wait.

The UN was held partly responsible for the problem. Large numbers of soldiers and foreign civilians contributed to an economic boom in prostitution and pimping, and fuelled an epidemic of STDs. In 1992, the rate of HIV among sex workers was estimated at 10%; by 1996, it had risen to 40%. At the same time, the average age of sex workers fell dramatically – by mid-1993 'virgins' were in high demand and available for less than $US20.[14] By the late 1990s, Cambodia's AIDS problem was considered the worst in Asia.

Detachment Medics

The Australian contingent, with its primary role of enabling communications, was dispersed across the country in detachments and border posts with Thailand, Laos and Vietnam.

---

[13] For an interesting insight into the extent of the STD problem in World War I, see Stanley, Peter, *Bad Characters,* Murdoch Books, 2010.

[14] Authors' note: Khmer prostitutes commonly worked from bamboo huts along the side of the road, screened only by a thin layer of fabric. Meanwhile other girls, with their trademark white-painted faces, would sit outside and chat. One of my less conventional jobs was to visit the brothels, chat with the girls and encourage them to use condoms with the soldiers. SN.

Each of the seven main detachments housed 12–15 Australians, usually with one or two women among them. Medical assistants at each detachment were required to operate semi-autonomously. In these pre-Internet days, their only link to the RMO was a single cumbersome 'mobile' phone (one of only three in the FCU) for advice, prescriptions or retrieval arrangements. There, medics were responsible for the health and hygiene of Australian and New Zealanders in the 'dets'; ensuring that antimalarial prophylaxis was being adhered to and responding to minor injuries and illnesses. They also became de-facto 'bush doctors' in the communities in which they worked – providing advice and dispensing medical assistance where they could – a situation that sometimes required skill and diplomacy.

Over the two years of the UNTAC commitment, a number of the medics deployed to Cambodia were also women – the most well-known was Sergeant Norma Hinchcliffe.[15]

Sergeant Norma Hinchcliffe CSM

'Storming Norma', as she was affectionately known, was an Amazonian woman posted as the RAP Sergeant in Phnom Penh. Awarded a Conspicuous Service Medal for her service in Cambodia, her experiences were also included in the *Australian Nurses at War* exhibition at the Australian War Memorial in 2012.[16]

In 1993 all newly enlisted medics were assigned to the RAAMC. Norma however, was a senior soldier with an established career in the Royal Australian Army Nursing Corps when she deployed to Cambodia. She proudly maintained her loyalty and refused to change her cap badge, or her allegiance. Nonetheless, her experiences are typical of medics at the time.[17]

---

[15] Both Sergeant Norma Hinchcliffe and Corporal Elizabeth Matthews – featured later in this chapter – were members of Susan Neuhaus's medical team.

[16] See 'Bandits in Battembang', *Willingly into the Fray: One Hundred Years of Army Nursing*, Ed. Catherine McCullagh (recalls Norma Hinchcliffe and Lew McLeod).

[17] The corps allegiance and accoutrements issues faced by female medics in the 1990s have remarkable similarity to those faced by the VADs/AAMWS in World War II.

Other people's wars: peacekeeping near and far

After a brief stint in Phnom Penh, Norma was transferred to oversee the largest of the medical detachments; in Battembang, a city imbued with an ancient past and the aftermath of genocide. On the outskirts, the bodies of thousands of Cambodians murdered by the Khmer Rouge had been dumped in 'killing caves'; on the streets, orphans were the legacy of systematic murder.

In Battembang, Norma took on a significant community role with street orphans, working with the G-11 club – an initiative of the first rotation of Australian and New Zealanders in conjunction with the local Catholic mission. They ran a delousing program, hired a teacher and ensured that the children had one hot meal a day from Monday to Friday.[18] Cambodia had become a nation of orphans and young beggars. Life on the streets was cheap, as Norma described:

> We tried to teach them basic hygiene, caring and sharing ... As we pulled up we would be welcomed with shrieks of delight and many little feet running towards our vehicle. They would fight over who was going to carry the first aid box, the bread and they would take our UN hats, which would be shared around. By the end of the day we would hope that we had not picked up any head lice. The kids became very capable at cleaning their cuts and knew they had to wash them before the Betadine went on, and then of course the band aid – we all know that the bigger the band aid, the better effect it has. The girls loved getting their hair washed and would sit in the sun to dry it. Sergeant Jodie Clark, the transport supervisor, and I made the fatal mistake of going and buying each of the girls a different-coloured outfit. Like all females, when we gave the outfits to them, they fought over them and basically ripped the clothes off each other. At the same time we bought the boys a shirt each, and by the week's end they had all sold or swapped them, so we didn't do that again.
>
> The saddest time of all with the street kids was when several girls who were about 14 years old disappeared. After

---

[18] Norma Hinchcliffe was awarded a Conspicuous Service Medal in recognition of her significant work with street children during the UNTAC mission.

we investigated, we discovered that a man had come along and wanted to marry one of the girls, but actually he had taken her and her girlfriends to be prostitutes in the Thai–Cambodian border camps.[19]

Determined to experience as much as possible during her deployment, Norma also had a penchant for misadventure. On 18 April 1993, she and Warrant Officer Class One Tim Hazeldene, the Health Officer, went on a routine trip to one of the outpost towns where Australian signallers were based, to check on living conditions and water supplies. They were travelling in a Bell 206 helicopter when, suddenly, they came under fire:

> We had left the village and were flying low over scrub when small-arms fire opened on us. We took a bullet through the windshield up into the console. The sound of the bullet hitting the console was deafening. It blew out all our comms and gauges. We had hydraulic fluid coming out and spilling onto us. Bob the pilot yells out, "They're shooting at us! They're f...ing shooting at us! Mayday! Mayday!"
>
> He turned and looked at me and said, "Are you OK? Have you been hit?" For about twenty seconds I couldn't answer as I thought I had been shot, as a piece of metal had hit me in my back. I put my hand to my back expecting to see blood. I said I was OK. Bob was yelling, "We have to land her or she will blow!" At the same time we were yelling for Tim through our comms and there was no answer.
>
> The pilot and I looked at each other with that look of horror and dread as we turned to see Tim sitting in the back (we had expected to see a bullet in his head). Thank God he was OK – he had only been hit with some metal.[20]

After a precarious landing and a long wait hoping for assistance, the helicopter and its passengers limped back to the nearby Dutch Marine Base. It was a close call.

---

[19] See: 'Bandits in Battembang', *Willingly into the Fray: One Hundred Years of Army Nursing*.
[20] ibid.

Other people's wars: peacekeeping near and far

Corporal Elizabeth (Liz) Matthews

Corporal Liz Matthews arrived in Cambodia in 1993, also as part of the second UNTAC contingent. Like Norma, she was also posted as a detachment medic – her base was in Siem Reap, home to the ancient temples of Angkor Wat at the northern end of Tonle–Sap, the inland waters in central Cambodia. Based four hours' drive from Phnom Penh, Liz knew that she would be isolated. She was warned that conditions could be hard:

> We did two weeks of intense training at Portsea to make sure that, as medics, we were ready to deploy in an environment that was potentially hazardous. We prepared for any medical situation that could occur, knowing that some of us would be in areas of isolation.
>
> I only spent a week in Phnom Penh. Then I went to Siem Reap where I was the main medic for 16 Australians and one

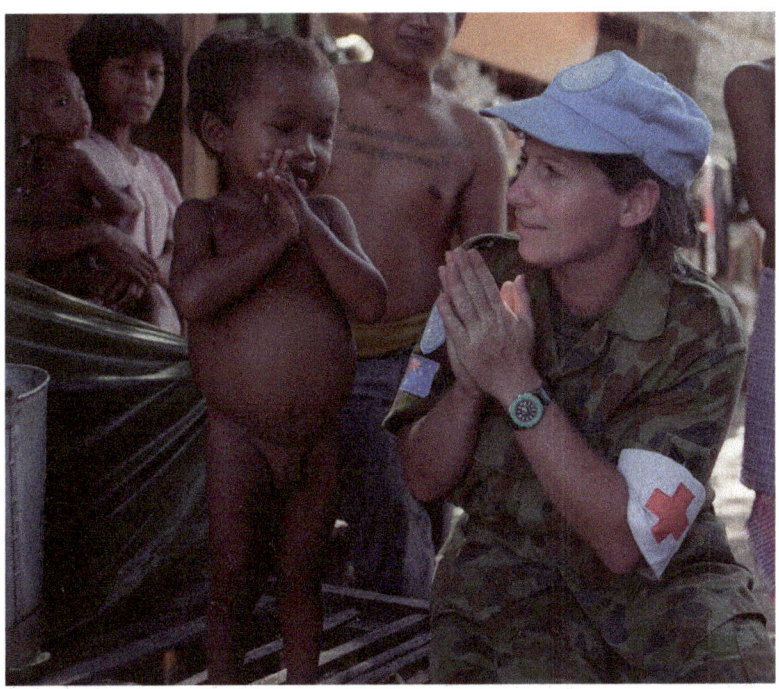

Corporal Elizabeth Matthews in Cambodia, 1993. *Image courtesy of Elizabeth Matthews*

Kiwi. I was the only medic in that area. There were a number of UN staff working with us, so I looked after them too. It was my first major deployment as a young corporal medic and it was exciting. It was daunting, but I knew I was ready.[21]

Originally from Victoria, Liz had joined the Army in 1986, shortly after her 21st birthday. She undertook recruit training at Kapooka – it began taking women only two years before she enlisted. "You have to look after yourself, look after your buddy, receive all that information and be militarized. The discipline is strict and some of the recruits struggled," she recalled of her time there.[22] Less than half of the seventy women who started with her at Kapooka made it through to the end.

Liz developed resilience during recruit training, but she also felt her age was an advantage. Unlike many of the other women who were only 17 or 18 years of age, Liz had 'life skills' – she had worked in hospitality as a barmaid and a cook; she had also embarked on a career in healthcare. When she arrived at the recruiting office in 1986, Liz had already decided that the daily grind of civilian life was not for her. "I was bored and I wanted an adventure. I wanted to be a medic and travel the world."[23]

Liz completed her initial medic training at Portsea and was posted to the 2nd Military Hospital at Ingleburn in Sydney. By the time she deployed to Cambodia, Liz had earnt a reputation for leadership and resourcefulness. She had shown she could work hard, and on her own. To prepare her for a role in a regional detachment, she was selected to undertake a three-month language course in basic Khmer at Point Cook near Melbourne. Although she was far from fluent, her ability to communicate with locals proved to be invaluable:

> There were two Cambodian girls who worked at our camp and I formed a very close bond with them. We used to buy our fresh fruit and vegetables from their parents' stall at the markets every

---

[21] Extract from interview between Sharon Mascall-Dare and Liz Matthews recorded 1 November 2013.
[22] ibid.
[23] ibid.

day. One of the things I treasured was that I could speak the language.[24]

Like other detachment medics, Liz too became involved in the local community. Together with some Irish civil policemen working in the district, she 'adopted' a Cambodian family; she helped to build a house; and she ensured that local children received adequate nutrition, while assisting with other small projects.

Like others who served in Cambodia and on peacekeeping operations since then, Liz also had to function in a multinational military environment. She worked closely with doctors from the Netherlands and from India – a challenging task given that some of the Indian staff were not used to working with female soldiers. She also spent time at the local Siem Reap hospital, working alongside volunteers from the multinational aid agency Médecins Sans Frontières. Cambodia's reconstruction was dependent on the support of the international community; Liz and her colleagues were constantly reminded of the realities of providing healthcare in a country starved of resources.

There was also a constant threat of violence. Like other regional centres, Siem Reap was a target for the Khmer Rouge, who were still pursuing a campaign of terror. "It was 'peacekeeping' in 'a sterile environment', but it wasn't," recalled Liz. "The Khmer Rouge was still raging around town, creating havoc. There was the potential of being shot. We still had to be on our toes."[25] It was a standing order that weapons, loaded with live ammunition, were always carried whenever outside the camp.

After eight months on the ground Liz returned to Australia. She went on to serve in Bougainville in 1994, in East Timor in 1999 and in Afghanistan from 2011 until 2012. Now a Warrant Officer Class One and senior medical technician for the RAAMC, Liz still remembers Cambodia as a key deployment in her career. "As a young female medic in the Army, and as a corporal, I received opportunities that I would cherish for the rest of my life."[26]

---

[24] ibid.

[25] ibid.

[26] Warrant Officer Class One Elizabeth Anne Matthews was awarded an OAM in the Queen's

Security and Electoral Violence

As Cambodia prepared to go to the polls in May 1993, UNTAC prepared for an increase in pre-election factional violence and tensions rose across the country. Security measures were increased and movements were restricted. Leaving the compound unarmed or without body armour was ruled out completely.

Sandbag barriers were erected around the RAP in Phnom Penh. The windows were taped to prevent the glass from shattering. Instead of morning physical training sessions, RAP staff helped to refill sandbags and repair trip-wired fencing. Sporadic gunfire in the vicinity of the compound was a daily occurrence, and the main catering depot to the UN, located less than a kilometre from the FCU, was under attack every night.

With the unit so widely dispersed in regional detachments, providing timely medical care to Australians in more remote areas was always a challenge. Even without factional violence, the landscape of Cambodia is hostile, marked by high plateaus, wide rivers and dense jungle. Road journeys were long and hazardous, and the three Perenti ambulances brought from Australia were unsuited to either the road conditions or the threat of landmines.

As a result, all medical inspections, detachment visits, evacuations and retrievals were conducted by the UN fleet of Russian MI-17 and MI-8 helicopters. Even then, air retrieval back to Phnom Penh could take significantly longer than the 'golden hour' and flying was dangerous, with flak jackets routinely laid on the aircraft floor as protection against bullets.

In the electoral lead-up, Australia reinforced its support to the UNTAC contingent and supplied three Blackhawk helicopters – the first time these aircraft had been deployed on operations. Repainted white with UN markings, they were sent to Battembang, with a small security detachment. With them came Major Carmel Van der Rijt, the first woman doctor in the Australian Army to have completed training in advanced aviation

---

Birthday Honour's List 2014: For meritorious service as the Health Operations Warrant Officer at Headquarters 17th Combat Service Support Brigade, the Health Warrant Officer at Headquarters Forces Command and as Army's Senior Medical Technician at the Directorate of Army Health.

Cambodia, 24 May 1993. Australian troops serving with the United Nations Transitional Authority in Cambodia (UNTAC) board an Army Black Hawk helicopter painted in the white of the United Nations. Left to right: Lieutenant Colonel Martin Studdert, Captain Susan Evans (back to camera) and Major Carmel Van De Rijt.
Image: George Gittoes. Courtesy of the Australian War Memorial, P01744.019.

medicine, undertaken during an exchange posting to Farnborough in the United Kingdom.

The Blackhawk Squadron provided a valuable retrieval capability at a time when it was difficult to access the north-western sector of the country. Tensions in the area were high as pockets of Khmer Rouge guerrillas carried out attacks operating from bases across the Thai border. Although there were no Australian casualties during their deployment, the team was scrambled to evacuate a critically injured Dutch soldier to Bangkok.

Flying at night, at low level across the Cambodia–Thai border, the Blackhawk arrived at Bangkok Airport unexpectedly and undetected by border radar. Although the casualty evacuation (or CASEVAC) was successful, it resulted in a diplomatic incident which Captain Evans had to explain at UN HQ.

Despite pre-election tensions, rumours and posturing, on 25 May 1993 the world was surprised. Confidence in the electoral process was overwhelming as 89.6% of registered voters turned out to decide their country's future. The relief was palpable; the pressure on UNTAC lifted temporarily. In the post-electoral 'haze', the FCU health team organised an event that is probably the UN's only 'Celeriter' competition, where military medical teams from different countries pitted their stretcher-bearing skills against each other. Teams from UN member states tested their speed, strength and agility in soaring 42-degree heat and drenching humidity as they traversed an obstacle course of unprecedented difficulty.

Despite proclamations of success, the results of the elections were not conclusive. FUNCINPEC (the royalist party) took 58 seats in the National Assembly, the Cambodian Peoples Party (KPP-representing the previous communist government) took 51 and the Buddhist Liberal Democratic Party (BLDP) took 10 seats. The result was a power-sharing arrangement between two Prime Ministers – Norodom Ranariddh and Prince Hun Sen. The scene was set for ongoing political turbulence that has continued to plague Cambodia.

With the elections over and a second Australian doctor in country, Captain Evans was allowed to take five days leave in June. She was in Australia when, on 21 June 1993, she heard the ABC news broadcast – her colleague, Major Susan Felsche, who was serving with the UN in the Western Sahara, had died in a plane crash. It was sobering news; a stark reality check for members of the RAAMC who had known Major Felsche and worked alongside her, and a reminder that 'peacekeeping' was not without risk.

The main body of the FCU returned to Australia in September and October 1993. A final detachment remained to escort Australian vehicles, ammunition and stores to Sianoukville, Cambodia's only deep-sea port, before the voyage home. After numerous delays, the odd security scare and several weeks on the white sands of Cambodia's coastline, Captain Peter Daniel, Captain Evans and a dozen soldiers said their goodbyes. They were the last Australians to leave the mission in Cambodia.[27]

---

[27] Twenty years ago after the end of the mission, Australian Defence Force members who served with the Force Communications Unit in Cambodia were awarded a Meritorious Unit Citation (MUC) in the 2014 Australia Day Honours List: for sustained outstanding service in warlike operations.

Other people's wars: peacekeeping near and far

Western Sahara

On 28 June 1993, Klaus Felsche stood at the front of Trinity Uniting Church at Wellington Point, south of Brisbane. "The words which follow may be the most important that I have ever and may ever say", he began. "They are about my best friend, my wife and my inspiration ... these words come from the heart".[28] It had taken Klaus many hours to write this eulogy for his wife's funeral; Susan Felsche was the first woman in the Australian Army to die on operations since World War II.

The week before, at a remote airfield in Western Sahara, Susan had died in a plane crash along with the pilot and one other passenger. Australia had lost a remarkable citizen and the Army had lost one of its best officers, Klaus explained, expressing his feelings of extreme disappointment that the couple would never share the life they had planned. He also hoped that his wife, and those who were with her, had not died in vain:

Major Susan Felsche c. 1993. *Image courtesy of Klaus Felsche*

> My heart also reaches out the families of those who perished in the accident with Sue. I know what they must feel and I hope that every effort will now be made to establish a lasting peace in the Western Sahara – too many well-meaning lives have been lost to win that elusive prize.[29]

Susan had been deployed to a conflict at the other end of the Earth, close to the antipodal point of her home town in Queensland. It was a conflict that would lead to one of the most protracted UN missions in peacekeeping history.

---

[28] Felsche, Klaus, Eulogy read at Susan Felsche's funeral, AWM Collection, PR00288.
[29] ibid.

Contested Territory

It has been called the nomad's no man's land – a remote expanse of desert, strewn with landmines in parts, that remains one of the most sparsely populated areas of the world[30]. Western Sahara is an inhospitable and inaccessible territory.

With the Atlantic Ocean to the west, Morocco to the north and Mauritania to the south and east, Western Sahara was a colony of Spain until 1975. Then, when the Spanish withdrew, Morocco and Mauritania moved in to stake their claims.[31] The local Sahrawi people resisted, forming a liberation movement called the POLISARIO front.[32] In the guerilla war that followed, Morocco turned to extreme measures to contain its enemy. In 1980, it began work on the 'berm' – a wall of sand and stone standing three metres high and running for 2,700 kilometers through the desert. Fortified with barbed wire, the wall was defended by artillery posts; it also had one of the highest densities of landmines in the world.[33]

The UN supported self-determination by the people of Western Sahara and the POLISARIO held out for a referendum, believing it had international support. In 1988, Morocco and the POLISARIO accepted a peace plan. The UN agreed to sponsor a ceasefire agreement and organise a referendum supported by a peacekeeping operation, the UN Mission for the Referendum in Western Sahara, or MINURSO.[34]

Originally, the mission was expected to include 800–1,000 UN civilians, 1,700 military personnel and a security unit of 300 police officers.[35] Australia agreed to provide a signals unit (a contingent of 45,

---

[30] Armstrong, Hannah, 'The Nomad's No Man's Land', *NYTimes.com*, 2013, Available at http://latitude.blogs.nytimes.com/2013/10/02/the-nomads-no-mans-land/ Accessed 10 October 2013.

[31] Mauritania withdrew its claim in 1979.

[32] POLISARIO is derived from the Spanish: *Frente Popular para la Liberación de Saguia el-Hamrou y de Rió de Oro.*

[33] See: Simanowitz, Simon, 'The Berlin Wall of the Desert', *New Internationalist*, 2009, http://newint.org/features/special/2009/11/10/sahara-berlin-wall/ Accessed 10 October 2013.

[34] MINURSO is an acronym from the French: *Mission Internationale des Nations Unies pour le Rèferendum au Sahara Occidental.*

[35] See http://www.un.org/en/peacekeeping/missions/minurso/background.shtml Accessed 10 October 2013.

reduced from 220, the number recommended for the mission[36]), including radio operators, drivers and medical support. The Australians were led by Lieutenant Colonel Gordon, a signaller, who had worked with a UN technical survey mission in Western Sahara in 1990.

Initially, progress on the mission was slow and frustrating. The Moroccans employed delaying tactics when they could and the Australians were frequently hampered by old and unreliable equipment. Eventually, living and working conditions for Australia's first contingent began to improve:

> Accommodation improved from the tents, Moroccan barracks and bombed-out Polisario buildings used in the initial stages. The distribution of fresh food improved, to supplement a diet which had been strong on locally sourced bread, rice, chicken, goat and camel. None of this brought the referendum any closer, and the deployment of Minurso began to stretch out indefinitely. After an eight-month tour, the first Australian contingent was replaced in May 1992; thereafter the Australians were rotated each May and November.[37]

In all, five contingents of Australians were deployed to Western Sahara between 1991 and 1994. Among their UN peers, the Australians had a reputation for professionalism and flexibility. In addition to their communications expertise they contributed convoy escorts, emergency response teams, welfare support, medical support and, importantly, the mission's most popular drinking hole – the Kangaroo Club.[38]

A number of dangers faced the Australians in the Western Sahara. Health risks included heat, malaria and food poisoning; much of the desert was heavily mined, making road transport precarious. Unpredictable winds played havoc with aircraft instruments and there were frequent days when no flights were permitted due to poor visibility.

---

[36] Londey, Peter, *Other People's Wars: A History of Australian Peacekeeping*, Allen and Unwin, Sydney, 2004, p. 140.

[37] Londey, Peter p. 142.

[38] Godfrey-Prendergast, Wilbur, 'The UN Mission to Western Sahara', *The Signalman* 1994, Vol. 28, p. 6.

Despite the dangers, all of the 219 Australians deployed to Western Sahara returned home safely, except one.

Major Susan Felsche

Born in Brisbane on 24 March 1961, Susan set her sights on a career in military medicine from a young age. She was active in the Naval Reserve as a medical student at the University of Queensland and by 1982 she had risen to the rank of Petty Officer. She was planning to join the Navy full-time, but changed tack in 1983 and joined the Army instead, "believing that it offered more challenging employment for medical officers".[39] In the early 1980s women were not allowed to serve at sea, limiting female naval officers to shore postings only. This contributed, no doubt, to Susan's decision.

A year later, in August 1988, she married Klaus Felsche, who was also an Army officer, and the couple settled in Canberra. Alongside her military career, Susan studied part-time and also worked after hours in the emergency departments of local hospitals to maintain her skills. At work Susan was dedicated to the job; at home she was Klaus's soul-mate – hard-working, down-to-earth and professional, she was also great fun to be around:

> She came from a normal Aussie family. All she did was with the support of her family and she had a strong service approach. She was very active in her youth: she was very involved in community event and issues. Intellectually, she was very bright and she was a very good doctor. She was very good at applied medicine. She also had a great sense of humour.[40]

In 1991, Susan was promoted to Major and posted to the Directorate General of Army Health Services. That same year she heard about the opportunity to deploy to Western Sahara. When she was offered a posting

---

[39] See http://www.awm.gov.au/people/1078604.asp Accessed 14 October 2013.
[40] Extract from interview between Sharon Mascall-Dare and Klaus Felsche conducted on 19 September 2013.

she accepted, keen to develop her skills in an operational context.

First, however, she was posted to the 1st Military Hospital at Yeronga in Queensland as the Medical Officer in Charge of Clinical Services; she also became a Fellow of the Royal Australian College of General Practitioners. By the time she began her pre-deployment training, Susan was well-established in her field of expertise – despite an interest in treating burns and trauma she chose to stay in general practice.

During her pre-deployment training at Randwick Barracks in Sydney, Susan wrote to Klaus several times a week, sharing her experiences of classroom lectures, weapons handling and fitness testing. Her letters conveyed her adventurous spirit and a sense of humour. She was unimpressed by the 'survival knives' presented during one briefing, telling Klaus they looked similar to the cutlery in their kitchen drawer at home.[41] Although weapons handling was not her forte, she persevered on the firing range and maintained her good humour:

> I wasted a few rounds on the lane marker before realising that I was supposed to be shooting at the thing that pops up. I did rather well … The positions were fun too – I just tried to take up the same position as the RSM in the lane beside me because I didn't have a clue what the terminology meant. However, I think I strained a few muscles in the process.[42]

Susan left Australia to join the 4th Australian Staff Contingent to MINURSO on 17 May 1993. The journey was long and circuitous, with stops in London, Madrid and Casablanca on the way to Laâyoune, the capital of Western Sahara. By this time the Australian contingent had successfully established communication networks from MINURSO headquarters in Laâyoune to UN sector headquarters around the country. There were also UN teams operating in various field locations, several hours' drive from the capital.

The Australians were responsible for driving supplies of fuel to UN

---

[41] Felsche, Susan, extract from letter to Klaus Felsche, 06 May 1993, AWM Collection, PR00288.
[42] Felsche, Susan, extract from letter to Klaus Felsche, 11 May 1993, AWM Collection, PR00288.

bases on either side of the 'berm'.[43] Susan had to pass a driving test in a UN 4 x 4 jeep shortly after her arrival; jeeps and light aircraft were the main modes of transport. "You would be proud of me," she wrote to Klaus. "I received my UN driver's licence ... I know where the engine is, spare wheel and jack."[44] Her test also required a hand-brake start up a steep, sandy slope driving the 4WD in reverse. It was no easy task.

Driving conditions, once the test was over, remained hazardous. The Sahara was an inhospitable place at the best of times, with relentless winds and shifting sands. The location of landmines could change without warning, marked by rocks that were visible one day but buried the next. The narrow roads could be treacherous – in one incident, Susan described how a Moroccan Army truck crashed head-on with a UN vehicle, killing one passenger and leaving three others seriously injured. With a damaged radio and no other means of communication, an Australian signaller further back in the UN convoy gave hand-written notes to passing vehicles to inform police about the incident. After a four-hour wait for help, he guided in a helicopter using a flashlight, marking the landing area with black tape.[45]

At first, Susan was based in Laâyoune. Her accommodation was comfortable, if basic, and she made the best of her surroundings:

> This place really is O.K. I have my own room and ensuite ... The paint is flaky and building quite old. It has a granite tile floor. The bathroom has a tiled floor, but it looks like someone put a very thin layer of concrete over it, splashed the walls a bit, then painted it. Shower worked well; loo was a problem. You pull the button to flush rather than push it. Of course I pushed it and the jolly thing disappeared for the night until I found my tweezers and fixed it. It's hard getting used to drinking bottled water.[46]

---

[43] See http://www.peacekeepers.asn.au/operations/MINURSO.htm Accessed 14 October 2013.

[44] Felsche, Susan, extract from letter to Klaus Felsche, 20 May 1993, AWM Collection, PR00288.

[45] Felsche, Susan, extract from letter to Klaus Felsche, 23 May 1993, AWM Collection, PR00288.

[46] Felsche, Susan, extract from letter to Klaus Felsche, 17–18 May 1993, AWM Collection,

Other people's wars: peacekeeping near and far

Susan was 'shown the ropes' by the Commanding Officer of the medical contingent in Western Sahara. On average the workload was low; she dealt with two inpatients and five outpatients a day although an outbreak of gastroenteritis increased the number of patients shortly after her arrival. One legacy of Spanish colonisation was a daily siesta, taken from noon until 3 pm every day. During those precious hours Susan wrote her letters to Klaus, and looked forward to the holiday the couple was planning to take in Europe at the end of her deployment.

Klaus warmly recalls the "youthful exuberance" of Susan's writing. From the outset she was keen to escape the capital and visit more remote areas of the country. "Hopefully I'll get out to Awsard or Smara where there is a Dr [sic] and nurse stationed who fly out to a different teamsite each day for a medical clinic … seems the best way to see the countryside," she wrote.[47] Awsard was a small town in remote, rugged country some 400 kilometres south of Laâyoune. The neighbouring UN base was little more than a camp of portable shelters surrounded by black rocky outcrops and sand.[48]

On 17 June, Susan arrived at Awsard for a rotation that was scheduled to last three weeks until 8 July. A nearby mountain offered spectacular views of the desert, but Susan was unimpressed by Awsard itself, describing it as a "military town with no women, crumbling stone houses half buried in sand and rubbish … generally an untidy, unhygienic type of place".[49]

From here, just west of the berm in Moroccan-occupied territory, Susan worked closely with the Swiss delegation to MINURSO, flying out to visit sick and injured members of the UN contingent and conduct clinics for local people. She was also confronted by a dilemma faced by RAAMC doctors before her and many to follow: "First one debates whether to help the POLISARIO at all, because then you have to help Moroccans," she wrote to Klaus. "They all like Western medicine and the whole business

---

PR00288.

[47] ibid.

[48] Londey, Peter, p. 142.

[49] Felsche, Susan, extract from letter to Klaus Felsche, 20 June 1993, AWM Collection, PR00288.

starts becoming untidy."[50] In one case, Susan's colleagues were faced with a POLISARIO baby that had fallen into a coma. The circumstances surrounding the baby's illness were unclear, and there was pressure to leave the POLISARIO to manage the case themselves. Still, Susan's colleagues did what they could and the baby survived.

The accommodation area comprised six 'Weatherhaven' shelters surrounded by a stone wall and a row of rocks marking the distant perimeter. Beyond the rocks were landmines – the area was out of bounds apart from a track up the nearby mountain that was carefully marked to avoid the mines on either side.[51] There were 20 bedrooms in the shelter and Susan was allocated a room at one end: "on two sides I have the outer wall of the tent, which flaps and makes a lot of noise," she wrote.

The next three days passed uneventfully. Susan spent her time seeing patients, cleaning her room and giving lectures on CASEVAC procedures. A highlight was the arrival of *The Australian* newspaper on Sunday 20 June; she also went into town that day with two friends. She wrote to Klaus, describing the restaurant that resembled an "old English barn" and the butchers, where "you chase the live chicken, they kill it and give it to you". It was the last letter she wrote.

The following day, 21 June 1993, Susan boarded a small Pilatus Porter aircraft at Awsard airfield. Within minutes of take-off the aircraft was in trouble and crashed, killing three of the four UN personnel on board. One of them was Susan. She was 32 years old.

Reaction to the accident was immediate. Her former CO at the 1st Military Hospital in Brisbane, Lieutenant Colonel Bob Millar, was quoted on the front page of *The Australian*, describing her as: "A spirited woman with a great love of life, a doctor protective of her patients and an officer with a shining career ahead of her." Her death received widespread coverage; she was the first Australian woman to die on an overseas operation since World War II. At her funeral, her family insisted on privacy – they had

---

[50] Felsche, Susan, extract from letter to Klaus Felsche, 17 June 1993, AWM Collection, PR00288.

[51] ibid.

lost Susan, a woman they loved, not the "female Digger" whose death had become a headline.[52]

Klaus also wanted answers. Frustrated by a lack of information and confusion about who was running the investigation into the accident, Klaus made the long journey to Western Sahara in 1994 to see the crash site for himself. "I believe that Australia has become much better at managing such processes," he says. "There have, no doubt, been benefits for the large and more frequent Australian peacekeeping operations since then."[53]

In all, 219 Australians served with the five contingents to MINURSO in Western Sahara. Retrospectively, the purpose of the mission has been reappraised. Its importance in establishing Australia's credentials as a key contributor of troops and expertise to UN peacekeeping missions throughout the world has been recognised.[54]

Susan's role, as part of Australia's contribution, is recorded in the peacekeeping gallery of the Australian War Memorial. A simple glass case contains her stethoscope, her photograph in a UN blue beret and Klaus's eulogy. "Like a lot of ordinary Australians she was doing her job," he says. "She was service oriented; she was very professional; it was part of who she was."[55]

In 1994, after fulfilling its role, Australia withdrew from Western Sahara. At the time of writing (2014), the UN mission is still ongoing. Although the ceasefire has largely held, skirmishes, disappearances and killings are still documented and a referendum is yet to take place.[56] From

---

[52] 'Female Digger Dies in Western Sahara Plane Crash', *The Australian*, 23 June 1993.

[53] Extract from interview between Sharon Mascall-Dare and Klaus Felsche conducted 19 September 2013.

[54] According to Professor David Horner, Australia's commitment to Western Sahara was closely linked to its intentions in Cambodia, with Australia trying to win support from the UN by demonstrating its willingness to help in countries outside of its area of immediate interest. See: Horner, David, 'Australian Peacekeeping and the New World Order', *Australian Peacekeeping*, Cambridge University Press, Melbourne, 2009, p. 54.

[55] Extract from interview between Sharon Mascall-Dare and Klaus Felsche conducted on 19 September 2013.

[56] Armstrong, Hannah, 'The Nomad's No Man's Land', *NYTimes.com*, 2013, Available at: http://latitude.blogs.nytimes.com/2013/10/02/the-nomads-no-mans-land/ Accessed 10 October 2013.

north to south, the berm still stands, dividing families and communities, scarring the landscape in one of the remotest corners of the world.

Somalia

Somalia, located on the horn of Africa, is Africa's easternmost country. In the early 1990s Australians watched as devastation and famine in Somalia and neighboring Ethiopia played out on nightly television news. Extensive media coverage was given to the work of CARE Australia, the humanitarian aid charity. It had high-profile supporters: Phoebe Fraser, the daughter of former Prime Minister Malcolm Fraser, was one of CARE's coordinators and the Frasers' involvement kept Somalia on the both the political and media agenda.

The roots of the Somalia conflict lay in the aftermath of colonisation. Once divided between Britain and Italy, Somalia became an independent state in 1960, falling under the leadership of Mohammed Siad Barre in the 1970s and 1980s. At first his military regime was supported by the Soviet Union; then the United States provided aid in return for access to strategically positioned naval bases at the entrance of the Red Sea.[57]

In the 1980s, Barre used the full force of his military against opposing clans in the north of the country. The US withdrew aid and the country collapsed into anarchy. Combined with drought and famine, the ensuing civil war left 300,000 people dead.[58] In January 1991, Barre was ousted from power and fled the country; leaving it in the grip of not just famine, but under the control of competing clans and factional warlords.

In April 1992, the UN established the United Nations Operation in Somalia I (UNOSOM I) to monitor the ceasefire in Mogadishu and escort deliveries of humanitarian aid. Later that year, in response to UN Security Council Resolution 794, the Australian Government committed to support humanitarian relief efforts.[59]

---

[57] See: Connor, p. 73.

[58] ibid.

[59] Australian War Memorial, 'United Nations Operation in Somalia (UNOSOM) 1992' Available at: http://www.awm.gov.au/units/unit_20244.asp Accessed 18 December 2013.

Other people's wars: peacekeeping near and far

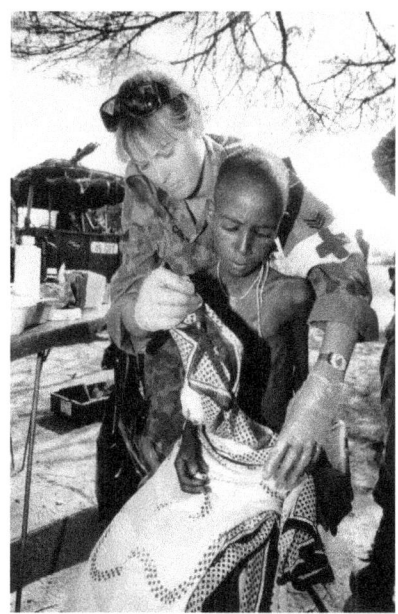

Sergeant Joanne Cook treating a Somali boy in Baidoa during Operation Solace.
Photo courtesy of Joanne Cook

The Unified Task Force – Somalia (UNITAF) was given a UN mandate to use "all necessary means" to perform its role.[60] Australia's contribution, called Operation SOLACE, comprised an infantry battalion group of 990 personnel, based around the 1st Royal Australian Regiment (1RAR).[61] Medical and other support was provided by the 3rd Brigade Administrative Support Battalion, which included seven female soldiers in its ranks.[62]

Corporal Kim Felmingham NSC

When Kim Felmingham was asked to go to Somalia she saw an opportunity to put her training into practice. Not only was it to be her first deployment,

---

[60] United Nations, 'Somalia – UNOSOM II – Mandat' Available at: http://www.un.org/en/peacekeeping/missions/past/unosom2mandate.html Accessed 13 June 2014.

[61] ibid.

[62] The six women were Sergeant Joanne Cook, Private Raelene Goldsmith, Private Christina Maclean, Private Lisa Lowe and Private Simone Steen. The final member of the group was Corporal (now Warrant Officer and Regimental Sergeant Major) Kim Felmingham.

Somalia, 29 March 1993. Corporal Kim Felmingham, Base Support Group, 1st Battalion, The Royal Australian Regiment (1RAR), serving with the Australian contingent to the Unified Task Force in Somalia (UNITAF). *Image: George Gittoes. Courtesy of the Australian War Memorial, P01735.467.*

it was her first trip overseas. Unfazed by what lay ahead, she felt ready to apply her skills in a warzone:

> You just practice, practice and then, when it finally comes, you switch into gear and you don't even have to think about it ... I had no medical exposure before I joined the Army, but you're trained and you put that training into practice. You do your exercise. You get deployed to Somalia. I don't mean to play it down, but it is pretty simple. You train hard, fight easy.[63]

Kim had joined the Army straight out of school, serving in the Royal Corps of Transport before transferring to the RAAMC in February 1991 at the age of 22. Her transfer coincided with increased flexibility regarding trades for women soldiers – Kim recalls that a number of nurses were also given the opportunity to transfer across to the RAAMC in the early 1990s. "Now we've got women in all trades and the opportunity is available to go into any trade, including all arms corps. It's open to everyone, regardless of gender, these days".[64]

---

[63] Extract from interview recorded with Kim Felmingham by Sharon Mascall-Dare 17 November 2013.
[64] ibid.

Other people's wars: peacekeeping near and far

The Australians were based in the Baidoa Humanitarian Relief Sector, west of Mogadishu. Their role was to maintain a secure environment in Baidoa itself; maintain a presence in the surrounding countryside; protect aid convoys; and assist in the equitable distribution of aid.[65] It was not an easy task as bandits loyal to local warlords sought to undermine the relief effort, attacking water supplies and stealing food. In one skirmish, an Australian patrol came across a group of bandits dismantling a water pump. In the firefight that followed, three Somalis were shot, a fourth was detained and several escaped.

Kim's role included escort tasks and building relationships with local people as well as medical duties. She saw herself as a "soldier first", drawn to the discipline and leadership that were intrinsic to the role.[66] As a medic, she was exposed to "everything from leprosy to traumatic amputations … anything you could possibly imagine".[67] She saw her first gunshot wounds as she dealt with coalition forces and locals who were casualties of the civil war.

One of her most memorable experiences was in the local marketplace, when she was required as an escort for intelligence forces:

> The intelligence soldiers were conducting patrols in the camel market in Baidoa and soldiers from the Battle Support Group [which included Kim] were utilised as security for the tasks. During the task, and in the middle of the market, it became apparent to the locals that I was female.
>
> The news travelled around the markets very quickly and in minutes we were surrounded by very curious locals. For the majority of locals, this would be the first time that they had seen a white woman, let alone a white woman in military uniform, carrying a weapon. Through interpreters, the intelligence soldiers established that the crowds were more curious than defensive.

---

[65] Australian War Memorial, 'United Nations Operation in Somalia (UNOSOM) 1992' Available at: http://www.awm.gov.au/units/unit_20244.asp Accessed 18 December 2013.

[66] Extract from interview recorded with Kim Felmingham by Sharon Mascall-Dare on 17 November 2013.

[67] ibid.

> I was able to talk to some of the local women, which was not allowed by the male soldiers for cultural reasons.
>
> The local women would give me signs of peace and clench[ed] fists as a sign of strength. It was very moving and rewarding to be able to witness such humility and hope in appalling and desperate conditions.[68]

In May 1993, the majority of Australian troops withdrew from Somalia, leaving a movement-control group and a small group of Australian air-traffic controllers. In October that year, 18 Americans were killed during the ferocious two-day Battle of Mogadishu. The brutality and horror of the Somali conflict, and the plight of the US Rangers, pinned down by sustained attacks by Somali militia, inspired the movie *Black Hawk Down*, released in 2001.

Kim went on to serve in East Timor, the Solomon Islands[69] and the Middle East, progressively moving up the ranks as she did so, continuing to make her mark as a soldier. Her deployment to the Solomons arose during her posting as Company Sergeant Major at the 2nd Health Support Battalion. In December 2012 she was appointed Regimental Sergeant Major of Joint Task Force 633 in the Middle East Area of Operations. The latter was a tribute to her leadership, skill and mastery as a soldier.

On 25 March 2000, she was awarded the Nursing Service Cross (NSC): "For outstanding devotion and competency in providing medical treatment to vehicle-accident casualties on 14 January 2000 while on Operation STABILISE in East Timor." Since the NSC was established in 1989, there have been 28 recipients and one bar; a number have been awarded to members of the RAAMC as well as the Royal Australian Army Nursing Corps.[70]

---

[68] Felmingham, Kim, Extract from private correspondence between Kim Felmingham and Sharon Mascall-Dare, dated 15 November 2013.

[69] Warrant Officer Kim Felmingham served in the Solomon Islands at the same time as Sergeant Kerry Summerscales and Captain Joanne Marks – both women are recognised later in this chapter.

[70] Past and present members of the RAAMC awarded the NSC include Corporal Warren Purse (1994), Corporal Leigh Wilson (1999), Sergeant Kim Felmingham (2000), Corporal Wayne McKenna (2000), Corporal Sarah Longshaw (2004), Captain Gregory Brown (2006),

Other people's wars: peacekeeping near and far

Kim has no doubt that her experiences in Somalia laid the foundation for her actions in East Timor:

> In Somalia we assisted with vehicle accidents, mass casualties and also accidents involving locals and US forces ... the time I spent in Somalia had an impact and assisted me with what I was doing in East Timor.
>
> Call it 'right place, right time' or not, you just kick in. You do what you've got to do.[71]

RWANDA

Amongst the paintings in the Australian War Memorial is a drawing by the war artist George Gittoes. Drawn at Kibeho, in Rwanda, in 1995, *Mass Grave* is more than mere art; it offers a unique perspective on a scene that challenges the very notion of 'peacekeeping' in the presence of war crimes.

In black and white, the viewer peers down a pit filled with the bodies of refugees, alongside four Australian peacekeepers wearing UN caps. "The young Australian soldiers at Kibeho would have seen more death than any Australian soldiers since World War II," said Gittoes in the notes that accompany the drawing. "They got to know the refugee families before the massacre – then saw many of them ruthlessly killed." Gittoes saw the evidence himself, firsthand, at Kibeho. He was present as Australian troops demonstrated extraordinary courage in the face of atrocity, and were powerless to prevent it; and he was witness to gallantry of a small medical team, led by a female Australian doctor.

Prior to the 1990s Rwanda was a relatively obscure and little-known African nation that was perhaps best known for its population of gorillas that had been studied in detail by the naturalist Dian Fossey – the subject of the 1988 film *Gorillas in the Mist*. In reality, Rwanda was a nation

---

Corporal Daniel Davidson (2007), Warrant Officer Class Two Alastair Mackenzie (2007), Corporal McQuilty Quirke (2007), Warrant Officer Class Two Geoffrey Cox (2007) and Corporal John Walter (2008).

[71] Extract from interview recorded with Kim Felmingham by Sharon Mascall-Dare on 17 November 2013.

'Mass grave' by George Gittoes. *Courtesy of the Australian War Memorial*, ART90437.

Captain Carol Vaughan-Evans tends to a wounded Rwandan in this drawing by George Gittoes. The artist's written observations surround the work. *Courtesy of the Australian War Memorial*, ART90439.

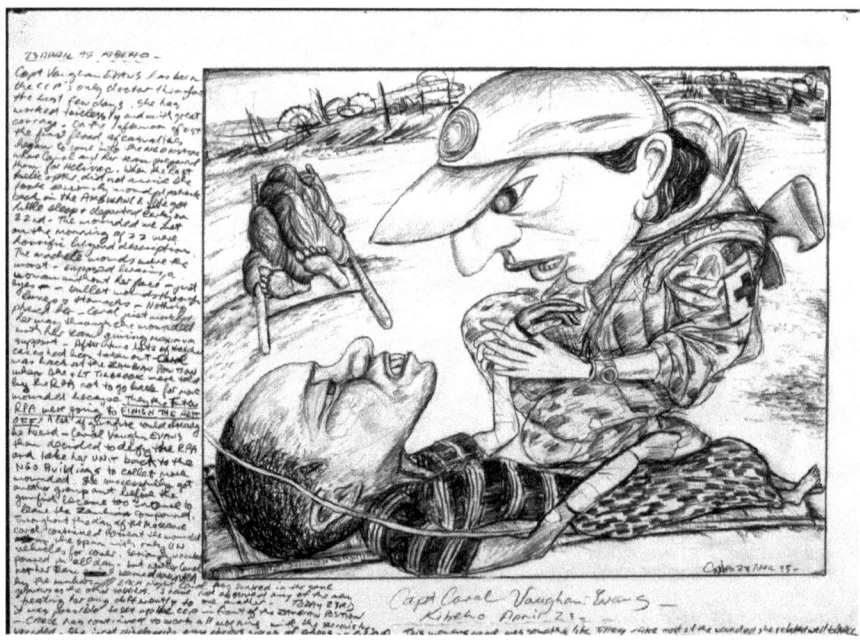

plagued by ethnic tensions between two ethnic groups – the Hutus and the Tutsis. In 1994, those tensions erupted.[72]

In the early 20th century, Rwanda was a German colony: it was part of German East Africa along with Tanzania and Burundi. When Germany ceded its Empire at the end of the Great War, Rwanda and Burundi became a trusteeship administered by Belgium, which exerted its colonial power to keep tensions between tribes under control. The country was divided between a minority Tutsi upper class and the majority Hutus, whose living standards continued to decline.[73] When the Belgian administration collapsed in 1959, the Hutu sought to wrest back control.

There were years of political instability. Rwanda's first President, post-independence, was Hutu; Tutsi rebels sought to overthrow his government. Tens of thousands of Rwandans fled riots and outbreaks of violence, only to encounter persecution across the border in neighboring Uganda. By 1994, the country had become a quagmire of inter-tribal conflict, displacement and chaos.

On 6 April 1994, a plane was shot down as it landed at Kigali airport. On board were the President of Rwanda and the newly elected President of Burundi. Both were killed. Suspicion fell on the Hutu Presidential Guard. The Rwandan President had offered concessions to the Tutsi-led Rwandan Patriotic Front (RPF) and Hutu hardliners were opposed to his efforts. On hearing the news, Tutsis prepared for the worst. As one survivor of the genocide recalled:

> When my father heard on the radio on 7 April that President Habyarimana had died in a plane crash the day before, he came to our room and said in a very weak voice, "Wake up … this is our last day and I am sure we will be in the grave before Habyarimana". Some of us started to cry at the thought of having to say goodbye to everyone in the family, but the little kids didn't

---

[72] The origins of Hutu-Tutsi ethnic conflict go back 500 years, when Tutsi cattle farmers migrated south from Ethiopia and established an ascendency in Rwanda, dominating the local Hutu. See O'Halloran, Kevin, *Rwanda: UNAMIR 1994/5*, Australian Military History Series – 1, Army History Unit, Canberra, 2012, p. 9.

[73] O'Halloran, p. 11.

understand. We all put on the strongest clothes and shoes we owned so we could be ready to run. But in which direction? We didn't know.[74]

What followed was genocide on an unprecedented scale. Organised mobs of Hutu launched a rampage of rape and murder that was systematic in its execution. One million people died in 100 days as grenades were thrown into churches and entire families were murdered by machete. There was no escape from the slaughter as Hutu hardliners rounded up village after village of Tutsi and moderate Hutu, killing thousands of unarmed men, women and children daily.

The world looked on in horror. It was an incomprehensible slaughter of Rwanda's own people; a stain on the lush mountainous country and a stain on the record of the UN, which, in response to increasing tension, withdrew its forces. Ten Belgian paratroopers serving with the UN were among those left unprotected by the UN withdrawal. They were murdered and mutilated by Hutu militia.[75]

Despite the genocide unleashed against their people, the Tutsi Rwandan Patriotic Army (RPA) mounted a successful offensive against Hutu government forces. By 19 July, the RPF and RPA had taken control of Kigali and other key strongholds and a tentative ceasefire was agreed. A new government, of Hutus and Tutsi, was established with a Hutu President and Prime Minister; a new post of Vice-President was created for Paul Kagame, a Tutsi who was also the Defence Minister.[76]

The UN responded with a rapidly assembled 15-nation Peacekeeping Force, the United Nations Assistance in Rwanda II (UNAMIR II). It had a different mandate from UNAMIR I, established by the UN the year before (in 1993) to monitor a ceasefire between the two parties.

---

[74] O'Halloran, Kevin, p. 39.

[75] The death of the 10 peacekeepers led to the withdrawal of the 450 Belgian troops in Rwanda and opened the way for the genocide to spread. On 5 July 2007, Bernard Ntuyahaga, a former Rwandan major, was found guilty of their manslaughter. The peacekeepers were brutally attacked, tortured and castrated. See Peterson, Scot. *Me Against My Brother: At War in Somalia, Sudan, and Rwanda: A Journalist Reports from the Battlefields of Africa*. New York and London: Routledge, 2000. p. 292

[76] In 2003, Kagame went on to become President of Rwanda.

Australia's contribution to UNAMIR II included 100 medical staff and a security force of around 200 Australian infantry and support personnel.[77] Their mission, under 'Operation Tamar' (Troops and Medical Aid Rwanda), was to provide medical support to United Nations forces. The mission was beset with complexity and difficulty. It was the first time that Australia had mounted a primarily medical mission with supporting infantry (rather than the other way round), and there was uncertainty about the mandate and the rules of medical engagement. Australian was also asked to provide humanitarian relief and medical assistance to international charity staff if they had 'spare capacity' – a capacity that would soon be overwhelmed.[78]

The first Australian contingent left for Rwanda in August 1994. Kigali Airport was dark and eerily quiet. They arrived to find the infrastructure of health services had been largely destroyed, since most Rwandan health professionals had either been killed or fled the country in the genocide.[79] The first weeks were spent cleaning a wing of the central Kigali hospital to make it suitable to accept patients. The walls were:

> … covered with caked-on blood and faeces and party members had to manually remove waste from the toilet bowls. Evidence of significant carnage was everywhere: patients had bolted leaving intravenous lines dangling, and the walls were spattered with the blood of those fleeing for their lives. The smell, alone, was stifling[80].

Six months later, in February 1995, a second rotation arrived. Despite UN intervention, and the formation of the new government in Rwanda, hundreds of thousands of internally displaced persons (IDPs) were still living in camps in southern Rwanda. At the height of the slaughter, some three million refugees had left their homes. The Hutu were unwilling to

---

[77] O'Halloran, p. 72.

[78] Miller, P., Pearn J., Marcollo, S. Radiology in Rwanda. *Australasian Radiology*, 1995:39, pp. 337–342.

[79] Smart, T. L., 'Medical Practice on the front line: separating the myths from the reality', *Medical Journal of Australia*, 2003; 179(11/12): 587–590.

[80] See: http://www.anzacday.org.au/history/peacekeeping/anecdotes/rwanda01.html Accessed 16 Jun 2014.

return, fearing revenge from Tutsi 'death squads' if they went home.

The largest of these camps was five hours by Landrover from the capital Kigali, near Kibeho, – a sprawling camp of makeshift shelters, housing between 80,000 and 100,000 people in a nine-square kilometre area.[81] Some sought sanctuary believing it was the site of miracles – in the 1980s, groups of children claimed to have seen visions of the Virgin Mary in a series of apparitions at a local church.[82]

On Tuesday 18 April 1994, the Australian headquarters was advised that the RPA had arrived at Kibeho, intending to shut down the camp. Around 2,000 RPA troops had closed in and refugees were flooding towards the two permanent UN Zambian positions on site.[83] There was suspicion that Hutu *génocidaires* had taken refuge in the camp – members of the Hutu militia who had participated in genocide. News of the shutdown took the UN by surprise. Australia was asked to provide a medical team to assist refugees immediately.

In the predawn light of 19 April 1994, 32 Australians left Kigali for the drive across rough terrain. The team comprised a medical section, a signals section and two sections of infantry to provide security. Their task was to provide humanitarian aid to the camp, but they had orders to seek permission from the RPA in order to do so. Their mandate was ambiguous in medical and military terms – if they encountered opposition from RPA troops, their ability to respond was limited:

> We had orders to cooperate with the Rwandan authorities and not to shoot at them, even if right under noses those forces were killing innocent civilians. Our guidance came from the Orders for Opening Fire cards. There was red card for opposed overseas deployment and a yellow card for unopposed overseas deployment. We were on neither card. [84]

---

[81] Milller, P., Pearn, J., Marcollo, S., 'Radiology in Rwanda', *Australasian Radiology* 1995: 39; 337–342.

[82] Pickard, T., 'Kibeho – Hell on Earth', *Combat Medic*, p. 55.

[83] O'Halloran, p. 110.

[84] O'Halloran, p. 3.

Captain Carol Vaughan-Evans MG

The Australian medical treatment team that went to Kibeho that morning was led by Captain Carol Vaughan-Evans, a relatively junior medical officer who had completed her medical training only a few years earlier. Aged 28, she was born in Johannesburg and had a family property in Zimbabwe, but nothing could have prepared her for the horrors she would encounter in Rwanda.

The team arrived to find that RPA troops had surrounded the camp and were herding refugees into an area the size of a football field on the ridgeline below.[85] There were abandoned shelters – some in flames – and personal belongings dumped on the side of the road.[86] As Carol described:

> The government forces had surrounded the camp and, in an effort to clear the locals out, had closed in very tight ranks, forced [sic] the people out of their homes basically denying them water, food and shelter in order to get them out. The locals didn't want to do that. They had fears about returning to their homes where they had lived in previously because returning there often there'd be squatters there, they'd be accused of murder and stoned.[87]

Immediately, the Australians set up a makeshift Casualty Clearing Post (CCP) next to the Zambian infantry company, the permanent UN force on site, putting up 'hootchies' to shelter casualties and using the back of an ambulance.[88] As the Medical Officer, Carol was responsible for the operation of the post, which included the collection, assessment and evacuation of casualties. Her actions, over the days to come, would later be cited for a Medal for Gallantry.

Tensions were clear the moment they arrived. Forced by the RPA to move a number of times, the Australians were unable to set up their

---

[85] Australian War Memorial. Wartime Issue 39 – Bravery under fire. John Connor. Available at: http://www.awm.gov.au/wartime/39/ Accessed 31 October 2011.

[86] Pickard, p. 56.

[87] Extract from interview with Carol Vaughan-Evans. Available at: http://www.australiansatwar.gov.au/throughmyeyes/transcripts/pk/cuf_vaughan_evansa.html Accessed 25 October 2013.

[88] O'Halloran, p. 131.

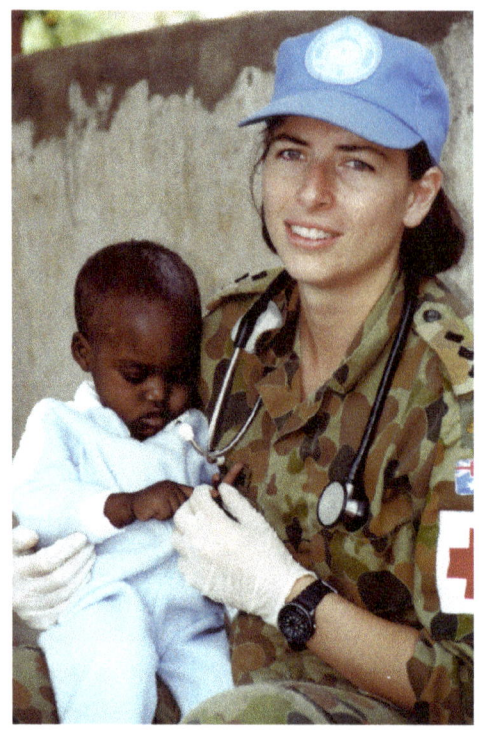

Captain Carol Vaughan Evans with Rwandan child. *Image courtesy of Major General John Pearn*

inflatable medical tents and had to work from the back of trucks and the ambulance. As the camp was cleared, the RPA insisted that the team only treat people who were injured and had decided to leave the camp. They were not allowed to treat anyone else. It was an invidious task: the only way the refugees could reach the Australians was by passing through a checkpoint where genocide survivors would identify those suspected of atrocities. Those identified were taken away by the RPA and, most likely, executed. Carol recalled:

> I remember getting there on the Wednesday and that was four days preceding the massacre and we certainly weren't wanted. The government forces made that very, very clear but did respect the UN's presence … they made our life difficult and insisted that we only treated people who had decided to leave the camp. We couldn't advertise. We had to have no more than two minutes

with people. We couldn't feed them. Very hard to explain what drugs to take and how to pick anything wrong when you're just watching people walk by.[89]

Over the next few days the refugees had to endure unbearable conditions, with no water, food or sanitation. Conditions continued to deteriorate and tensions within the camp reached boiling point. On the morning of 22 April, after retreating to the relative safety of the Zambian military compound the night before, the Australians returned to the camp to find that many refugees has been killed or wounded during the night. Half had gunshot wounds caused by Rwandan soldiers; the others had machete wounds from the Hutu militia – they were trying to terrorise refugees into staying in the camp to provide a 'human shield'.[90] Sporadic gunfire continued through the day and the Australians took refuge in the Zambian compound at night, returning to the camp the following morning through increasingly hostile RPA checkpoints.

The medical team continued to treat as many people as it could. The humanitarian agency Médicins San Frontières (MSF) had established a small hospital adjacent to the camp and was treating large numbers of casualties – victims of either bullets or machetes, many with machete wounds that had cleaved their faces or hands – classic 'defensive injuries'.

The MSF hospital was overwhelmed and Carol and her team did what they could to help. Patients were triaged, treated and where possible evacuated by helicopter or by road. As evidence emerged that the RPA were trying to hide the number of killings, Australian infantry soldiers acted as stretcher-bearers, collecting as many casualties as they could, including children either trampled or macheted. Amid confusion and chaos, there was growing intimidation by RPA forces:

> The government forces were extremely aggressive, indicating that if we didn't empty the hospital they would, and their means were one of going through and killing people who remained.

---

[89] Extract from interview with Carol Vaughan-Evans. Available at: http://www.australiansatwar.gov.au/throughmyeyes/transcripts/pk/cuf_vaughan_evansa.html Accessed 25 October 2013.

[90] Connor, John, Bravery under fire, Wartime, AWM, issue 39. Available at: http://www.awm.gov.au/wartime/39/bravery/ Accessed 25 October 2013.

> So we returned as many as we could who were walking wounded back to within the confines of the camp itself and took those that we could treat and that was, we took everybody. We left no-one to have the government forces go through. The mood deteriorated further and whilst we were working on these people, people within the camp decided to run for their lives.[91]

That morning, Rwandan troops moved into the MSF hospital and started shooting. Soon afterwards, in heavy rain and thick mud, refugees began to stampede in a desperate attempt to break out of the camp. Some managed to surge over the razor wire into the Zambian compound to be met by Australian soldiers with bayonets fixed – the first time Australian infantrymen had done so since the Vietnam War.[92] Others were not so fortunate and fell victim to the RPA, who took up positions on the spur line and opened fire with automatic weapons, rocket-propelled grenades and a 50-calibre machine gun. Those who escaped were hunted down in the valley and executed:

> The people within the camp decided to run for their lives. I don't actually know what precipitated it but they ran through the human barricade of government forces and I guess the more they tried the better the success ... so lots of people ran. They didn't succeed. They were shot, bayoneted and the government forces chased them down. That went on for some period of time. There was a storm at the same time so it really was quite a tumultuous event.[93]

Throughout the carnage the Australian infantry held their positions, restricted by their rules of engagement and the UN mandate, which prevented them from returning fire. They were continually provoked by the RPA, who tried to force a response by executing men and women in

---

[91] Extract from interview with Carol Vaughan Evans. Available at: http://www.australiansatwar.gov.au/throughmyeyes/transcripts/pk/cuf_vaughan_evansc.html Accessed 25 October 2013.

[92] See: Connor, 'Bravery under fire'.

[93] Extract from interview with Carol Vaughan-Evans. Available at: http://www.australiansatwar.gov.au/throughmyeyes/transcripts/pk/cuf_vaughan_evansd.html Accessed 25 October 2013.

front of them and taunting them with atrocities. As one soldier described:

> They tend to pick on people that wouldn't be of much benefit in a stampede – small children, slow-moving pregnant women, the elderly. And just slowly mutilate them to put fear into these people. They would grab groups of 100 or 200 and stampede them and try to make a break through the lines.[94]

Although they felt powerless, in the face of such atrocities, the Australians' restraint undoubtedly saved their lives that day.[95] Nonetheless it was a horror for which no amount of training could have prepared them; it would haunt them many for years to come.

Meanwhile, Carol and the medical team continued to work from a makeshift casualty clearing post, working out of the back of an ambulance. All were weighed down by flak jackets and helmets; the stifling African heat and a rain downpour adding to their difficulties. Protected only by a flimsy sandbag wall and with automatic fire continuing to pepper the area around them, the medical team remained focused on their task. The danger and stress of the working environment was compounded by the fact that many of the Rwandans were HIV positive or had full-blown AIDS, at a time when no effective treatment was available and transmission of the disease was a death sentence.[96]

The team worked tirelessly to triage the injured and save who they could. As medical supplies became low, they retrieved what they could from the bodies, reusing bandages and even intravenous cannulas. The RPA tried to stop Carol using stretchers to remove casualties as they wanted to finish them off, but she steadfastly and defiantly ignored them, continuing to sort, prioritise, treat and evacuate whoever she could.

By about 1600, still in continuing rain, the first casualties were loaded

---

[94] Extract from interview with Robbie Lucas in 'Remembering the Forgotten Diggers', broadcast on *Sunday*, Channel Nine, 25 September 2005, transcript available at http://www.peacekeepers.asn.au/operations/UNAMIR%20more.htm Accessed 25 October 2013.

[95] See: Jordan, Paul, 'Witness to genocide: a personal account of the 1995 Kibeho massacre,' *Australian Army Journal*, Vol. 1, no. 1, June 2003, pp.127-136. Available at: http://www.anzacday.org.au/history/peacekeeping/anecdotes/kibeho.html Accessed 25 October 2011.

[96] Smart.

into UN helicopters for evacuation to Kigali. In a last effort, Trooper Jon Church, a medic with the Australian SAS, who was later killed in a Blackhawk helicopter training accident in 1996, decided to rescue a three-year old girl he found crying and wandering alone through the debris and carnage of the massacre. Knowing that their vehicle would be stopped at checkpoints and searched by the RPA, another medic bandaged her arm, pretending that she was injured, fed her a biscuit soaked in sedative and hid her in the ambulance blanket rack. After being held up at RPA roadblocks, the little girl was handed over to the Butare hospital and taken to an orphanage.[97] As Carol later wrote:

> We always remember that as a small victory. Despite all the [Rwandan army] did to that mass of humanity, we got that one little girl out of there.[98]

As Australia's official war artist, George Gittoes had accompanied the team to Kibeho. He witnessed the courage of Carol and her team, seemingly oblivious to danger, as they did what they could, directly in the line of fire. The team repeatedly ventured back to the MSF hospital to retrieve casualties even though they were repeatedly warned by the RPA to stop or be killed.[99]

> I saw moments of incredible human courage from the Australian peacekeepers. There was one moment where the Australian section was out, they'd gathered wounded and they came under fire from people within the refugee compound and it's terribly hard for any human being when bullets are fired at them not to try and take cover and protect themselves …
>
> There was also a moment where Carol Vaughan-Evans was told that we couldn't get back to get the people who had been bandaged because the RPA were going to finish them off, and a lot of heavy fire had started and we all looked to her and she made the decision to go back, which took tremendous courage.

---

[97] See: Jordan, P.
[98] See: Connor, 'Bravery under fire'.
[99] Pickard, T., 'Kibeho – Hell on earth', *Combat Medic*, p. 115.

Other people's wars: peacekeeping near and far

> It was one of the most courageous acts I've ever seen because she wasn't just risking her life, she was risking the lives of the people that were working under her. But you could tell that there wasn't even a fraction of a moment where she would have considered not going back for those wounded.[100]

Early the following morning, on 23 April, the Australian team re-entered the camp to conduct a search for remaining casualties and count the dead. This gruesome task revealed mothers killed with babies still strapped to their backs and children listlessly sitting in piles of rubbish and corpses.[101] The numbers killed have been widely contested by Rwandan authorities but the Australians counted over 4,000 dead before they were prevented from continuing. Other estimates were of 25,000 killed.

By now, more staff had arrived to reinforce the medical team. The 650 remaining wounded were triaged, treated and evacuated as necessary. Exhausted, Carol and her team had done what they could. Although they wanted to stay and continue working, they were ordered back to the main hospital. There, for the first time in five days, they could shower, wash and fall into an exhausted sleep.

Following the events at Kibeho, four Australians were awarded the Medal for Gallantry for their distinguished acts of gallantry in action in hazardous circumstances.[102] One of them was Captain Carol Vaughan-Evans:

> As the Medical Officer in Command, Major Vaughan-Evans was responsible for the operation of the Casualty Clearing post, which included the collection, assessment, and evacuation of over 500 severely wounded casualties ...
>
> Major Vaughan-Evans continued to treat casualties despite the risk of personal injury. On numerous occasions Major Vaughan-Evans accompanied casualties to the helicopter landing

---

[100] Extract from interview with George Gittoes, Available at: http://www.australiansatwar.gov.au/throughmyeyes/transcripts/pk/cuf_gittoes.html Accessed 14 June 2014.

[101] See: Jordan, P.

[102] The four Australians were Corporal Andy Miller, Warrant Officer Rod Scott, Lieutenant Steve Tillbrook and Captain Carol Vaughan-Evans. It was the first time Australian soldiers had been awarded gallantry medals since the Vietnam War.

zone while firing was still occurring in and around the camp. Throughout the entire crisis, Major Vaughan-Evans displayed acts of gallantry, inspirational leadership and exceptional medical support. Major Vaughan-Evans was undaunted by the hostile rifle fire and mass casualty situation which confronted her.

By her gallant performance of duty, distinguished leadership and tireless and selfless efforts, often under fire and always under appalling conditions, Major Vaughan-Evans was directly responsible for the saving of many Rwandan people. Her calmness in this life-threatening situation and her ability to make clear and accurate medical assessments under pressure were of the highest order. In addition her compassion and dedication to those she was treating, ability to improvise when supplies ran low, and outstanding medical expertise were in the finest traditions of the Royal Australian Army Medical Corps. Her acts of gallantry and leadership whilst under fire were inspirational to all members of the Australian Medical Support Force team at Kibeho.[103]

At the time of writing (2013) Carol Vaughan-Evans remains the only woman in the history of the Australian Army to have received such an honour.

Bougainville

The copper-rich islands of Bougainville sit like jewels in the Pacific Ocean. Yet for nine years from 1988 to 1997, the islands' natural beauty was ravaged by one of the most destructive and violent conflicts seen in the Asia Pacific region.

Bougainville's turbulent history has its origins in an aberration of 19th century cartography. Then, under German rule, these richly resourced islands were considered part of New Guinea, even though they are located some 1,000 km east of Port Moresby. Ethnographically, the people of Bougainville are Melanesian; they share a geographic, linguistic and cultural

---

[103] Excerpt taken from full citation. Available at: http://www.raamc.org.au/web/index.php?History:For_Valour Accessed 14 June 2014.

bond with the Solomon Islanders. Their separation led to division, with fractured lines of tribal and cultural misunderstandings and destructive consequences.[104]

Conflict over the Panguna copper mine[105] on the main island of Bougainville amplified such grievances – the economic and environmental impacts of mining operations were divisive.[106] The local people received little financial benefit from one of the world's largest open-cut mines, and witnessed dramatic changes to local ecology and demography:

> For villagers the technology of helicopters, drills, 105-tonne trucks and bulldozers ravaging their luxuriously verdant mountains was terrifying. There were also some 10,000 construction workers, nearly all male, mostly alien and seemingly, menacing. A billion tonnes of ore was eventually to be processed; the crater left would be four square kilometres: the Jaba River would be polluted for 50 years.[107]

In late 1988, the Bougainville Revolutionary Army (BRA), led by Francis Ona, embarked on a sabotage campaign against the mine. What followed was escalating conflict, demands of autonomy and guerrilla warfare that forced the mine to cease operations. The Papua New Guinea Police and then the Papua New Guinea Defence Force (PNGDF) were deployed to squash the secessionist rebellion. Their efforts resulted in violent reprisals, summary executions, a military blockade and over 70,000 Bougainvilleans displaced into 'care centres', isolated from the rest of the world.

In October 1997, the parties agreed to a truce and the New Zealand-led

---

[104] Regan, Anthony, J., 'Melanesia: The Future of Tradition', *Cultural Survival Quarterly*, issue: 26.3 (Autumn 2002) Available at: http://www.culturalsurvival.org/ourpublications/csq/article/bougainville-beyond-survival Accessed 14 June 2014.

[105] The Bougainville Copper Mine, operated by a subsidiary of Conzinc Rio Tinto Australia (CRA), was one of Rio Tinto's most profitable operations.

[106] Regan.

[107] Criffin, J. (1995), *Bougainville a challenge for the Churches*, Catholic Social Justice Series, No. 26, 12. Quoted in Nicholson, S. and Hooper, H., 'Living up to the Pacific Promise – New Zealand's role in peace-building in our neighbourhood', in Tie, W. (Ed.) *Just peace? Peacemaking and peace building for the new millennium*, Conference proceeding, Auckland: Massey University, 2000.

Truce Monitoring Croup (TMG), comprising 50 monitors from Australia, New Zealand, Fiji and Vanuatu, was deployed. On 30 April 1998, the parties signed a permanent ceasefire agreement. The Australian-led Peace Monitoring Group (PMG) then replaced the TMG.

At its peak, the PMG comprised about 300 personnel.[108] Protected only by canary-yellow baseball caps and armbands with a white dove against an azure background, Operation Belisi[109] marked the beginning of a new era in Australian peacekeeping. It was a multiagency, diplomatic partnership where Australian soldiers, diplomats, police and civil servants worked together to reach a sustainable political solution for our neighbours in the Asia Pacific region. [110]

In July of that year, Major Susan Neuhaus was on leave, at the Military Club, Cercle National des Armées, near Boulevard Haussmann in Paris, when she was asked to consider a deployment to Bougainville. Now a Reservist and Surgical Registrar, the decision would depend on her colleagues in the Department of Surgery, where she was working as a researcher, accepting her absence for several months.[111]

The Combined Health Element (CHE) comprised 20 personnel and was effectively a small field hospital, operating under canvas within the copra-lined sheds of the PMG's main support base at Loloho Wharf.[112]

There was a single operating theatre using a modified NATO litter stretcher as the operating table, two ventilated ICU beds and a six-bed ward. The pathology capability included a water-testing and environmental-health element; X-ray facilities; a resuscitation bay; a dental unit and a pharmacy. The CHE provided healthcare to the five main team sites and villages dispersed across rugged mountainous terrain of over 900 square

---

[108] In total, around 3,800 Australian Defence Force personnel and 300 Australian civilians served at various times with the PMG.

[109] Bel Isi is Tok Pisin for 'Peaceful'.

[110] Breen, B., 'Towards Regional Neighbourhood Watch', *Australian Peacekeeping: Sixty Years in the Field*, Ed. Horner, D., Londey, P., and Bou, J., Cambridge University Press, Cambridge, 2009, Chapter 4.

[111] See: Epilogue for a fuller account.

[112] The role of the CHE was to provide health support to members of the PMG. It was authorised to use any 'spare capacity' for emergency humanitarian or life-saving aid.

kilometres – much was accessible only by helicopter.

A specialist surgeon and anaesthetist were flown in every two to four weeks throughout the mission. Much of the cohesion of the team depended on the specialists, and towards the end of journey back from Aropa airfield with the new cargo, Warrant Officer Class Two Guiliani, the team's senior soldier, would utter a code word: "caramel crème/almond brittle …" summing up their personalities – he was invariably never far wrong. Only a few days before leaving Major Neuhaus discovered that there would be no general surgeon deployed during her tenure as the Officer Commanding – "We thought you could cover that", was the comment from the Director of Medical Services. It was a less than ideal situation, which, thankfully, was not repeated.

Over the three years of the CHE's operations, women were employed in almost all roles. They served as commanders, clinicians, medical assistants and support staff. Drawn from all three services, their primary role was to provide emergency and surgical level-health capability for members of the PMG. In reality, over 95% of the workload involved emergency care for local Bougainvilleans.[113]

In mid-1998, during visits to the local health clinic, Major Neuhaus and Flight Lieutenant Sindy Vrancic, a RAAF medical officer, had been monitoring the health of a young local woman called Nellie; her pregnancy was complicated by placental malaria and so the team were supporting her progress.

One morning, on arriving at the clinic Nellie was found to have delivered her baby through the night; gravely premature, the baby boy (later nicknamed Peter Rabbit) did not have the strength to suck properly and was dangerously underweight. Nellie and her baby were transferred to the CHE where the team demonstrated the remarkable ingenuity that is so characteristic of Australian soldiers. From little more than a polystyrene fruit box, an upturned walking frame, two fluorescent lights and bubble wrap, they constructed a makeshift humidicrib. 'Peter' was fed through a fine tube inserted through his nose into his stomach until he showed signs

---

[113] Neuhaus, S. J. 'Post Vietnam – Three Decades of Australian Military Surgery', *ADF Health*, 2004:5:16–21.

of gaining weight. Nellie herself was dangerously frail, not just from her recent delivery, but from the long-term effects of malaria ravaging her liver and spleen.

Once strong enough, he and Nellie were transferred to the local hospital on the northern island of Buka. Flight Lieutenant Vrancic bundled up the baby in her arms for the helicopter journey. Lulled to sleep by the vibrations of the rotor he slept safely in her arms through the helicopter journey. Weeks later, now much stronger, Nellie and her baby returned to Arawa and went home to their village.

During the peak operation of the Panguna mine, Arawa had a hospital and nursing school, which was unrivalled in the Pacific. Despite the hospital being completely destroyed and without equipment or medical supplies, their midwifery skills had survived the decade of the blockade; what they knew had been passed down through word of mouth with great memory and precision.[114]

Despite their experience and skill, the nurses were shy women. When asked for an opinion or a decision they were reluctant to speak, but once trust was established their eyes would shine – their barefoot smiles all the tacit approval that was needed.

Throughout the early years of the CHE the most common operation performed was a Caesarian section.[115] This was, in part, because the midwives brought their most difficult cases to the CHE or because for most of the surgeons – unfamiliar with obstetrics – it was the safest option in their hands. It is a tribute to the medical teams on site that the infant-maternal mortality rate from such interventions in the CHE was zero.[116]

Trauma presentations were also common. Often inflamed by consumption of 'home bru' (a dangerous concoction of illegal home-made alcohol), machete wounds were the signature of old grievances under a system of 'payback' which had deep cultural roots.[117] Frequently machetes

---

[114] A profoundly moving example of this gifted ability was evident each Sunday at Catholic Mass in Arawa; women would singing the Angelus in exquisite four-part harmony – without sheet music: the notes preserved only by faith and memory.

[115] Neuhaus, S. J. Post Vietnam.

[116] ibid.

[117] Availability of 'Home Bru' was a significant problem on the islands. Prohibition of alcohol

would fracture bone, resulting in compound fractures that needed to be scrubbed clean of jungle detritus to avoid infection or non-healing. Severe traumas were uncommon, but tested the team's resources to its limits.

These were a resilient people, but there was also another factor in the survival rates – the availability of fresh whole blood via so-called 'hot transfusion'. Whilst a fridge was kept stocked with temperature-controlled blood sourced from the Australian Red Cross, it was a scarce resource, reserved for members of the PMG. When blood was needed for a local, it came from willing members of the PNG or from family members.

Once considered inferior to concentrated blood products, the enthusiasm for whole blood has in recent years undergone resurgence.[118] Modern trauma surgery emphasises that deaths in trauma are not only due to injuries, but the secondary 'lethal triad' of loss of fluid, hypothermia (cold) and failure of the body's clotting system – warm, fresh blood provided an ideal solution.

Corporal Kerry Summerscales CSM

Throughout the CHE's operation, surgeons were reliant, not just on the expertise of the nursing staff, but also on the availability of X-rays, test results and blood. Many medical technicians from each of the services provided this support. One of them, Corporal Kerry Summerscales, was recognised for her efforts with the award of a Conspicuous Service Medal.

Kerry Summerscales arrived in Bougainville in April 1999, to an initially daunting environment. Recently promoted to the rank of Corporal, it was her first overseas deployment. She arrived on a Hercules and set off for her base at Loloho near the capital Arawa for a brief handover:

> It was all new; it was nerve-wracking. It was hot, it was humid and there was definitely an aroma that was a bit different. We drove from where we landed … and there was jungle like you've

---

was strictly enforced.

[118] More accurately it is provision of all the components of blood in a 1:1:1 ratio that has been associated with improved survival rates, particularly during Operations Desert Storm and Iraqi Freedom. This practice is now reflected in the Australian Red Cross' massive transfusion policy.

Corporal Kerry Summerscales talking with locals in the Solomons Islands.
Image courtesy of Kerry Summerscales

never seen before. It took over everything and the buildings were all covered in this jungle. I remember looking out the back of the vehicle and seeing a guy standing there with one leg bent up and I thought, "Oh, God! He looks evil!" Because he was so strong and masculine and muscle-bound; there's no fat on these people. And then he smiled. They've just got the most amazing smile you've ever seen in your life.[119]

Kerry joined the Army at the age of 18, in May 1990. Outgoing and adventurous with a sharp sense of humor and wicked laugh, Kerry also had a strong sense of justice. As a teenager, she had joined Amnesty International and was committed to human rights: "The reason I believe in [sic] Army is because we defend those who can't fend for themselves," she explained. "My main reason for going overseas is because we helped

---

[119] Extract from interview by Sharon Mascall-Dare with Kerry Summerscales on 21 October 2013.

the men, the young girls and the women and the young kids have some kind of future".[120]

While her parents were initially shocked by her choice of career, they soon realised that their daughter had an aptitude for military life. Standing only 153cm tall, Kerry struggled at first to handle the self-loading rifle (SLR), the standard weapon used in recruit training, but grew in confidence as the weeks passed. "There's things you never thought you'd do, like, what did I know about firing a weapon?" she recalled. "The SLR is 114 cm long, I'm merely 153. It was huge. I thought, 'I can't fire that.' But you do. You develop a confidence and a sense of pride and sense of bearing".[121]

After two and a half years as a medical assistant, Kerry decided on a trade-transfer to the Army's pathology technician stream. In 1997, she began two years of long-term schooling at the Queensland University of Technology, graduating with an Associate Diploma. She was promoted to the rank of Corporal and posted to Lavarack Barracks Medical Centre.

In Bougainville, Kerry soon found her training as a laboratory and pathology technician put to the test. As part of the PMG's 'goodwill' role, the Australians were sometimes required to treat local chiefs and dignitaries. Kerry had the opportunity to meet Joseph Kabui, later the first President of Bougainville after independence. Her memories, however, focus on the local people she encountered, and scenarios that required her to demonstrate speed and professionalism under pressure.

She became known for her skills in 'hot cross-matching' where she would conduct tests on blood, just taken from a donor, to determine its safety and suitability for transfusion into a patient who urgently needed it. The process of testing and analysis required total focus and concentration, under time pressure, to save a patient's life – a technique little changed from World War II:

> It's called a 'hot cross-match' because the blood is 'hot', straight from one body to another … and there's potential for it to go wrong. I was concentrating, and people would talk to me, and I would say "Ssh! don't talk to me, don't talk to me!"

---

[120] ibid.

[121] ibid.

> You're writing everything down and making sure you have the reactions recorded. It's such a precise thing. Normally, in the laboratory, you send the blood off and you don't see the result. In Bougainville the patient was 10 metres away from me so I got to see the result. He lived. He stopped bleeding.[122]

Like others in the CHE team, Kerry also worked outside of her area of expertise, providing training for local nurses and treating women who had experienced obstructed labour during childbirth. She remembers being "horrified" by stories of women walking for two days to the hospital with dead babies inside them, and feeling embarrassed after she assisted a midwife to deliver a woman's baby.

As her deployment continued, Kerry became more proficient in Tok Pisin and used her language skills to engage with local women. As well as pregnancy and sexual health issues, she was also exposed to domestic violence and reprisal 'payback' attacks on women.

There was no question in Kerry's mind that Australia's medical team had an important role as part of the PMG:

> We were keeping the people who were key players healthy and we were looking after the community; what's the point in maintaining a peacekeeping process if you've got no community?
>
> I just love Bougainville; it's a beautiful place. It's a shame that such violence was there. I'm horrified that we played some part in the history of that violence. But I'm also very proud of our history in trying to mend that.[123]

On her return from Bougainville, Kerry was awarded a Conspicuous Service Medal in the 2001 Queen's Birthday Honours. The citation recognised her hard work and positive attitude and also praised her efforts in encouraging others to maintain their fitness with physical training (PT) programs. Above all, she was recognised for her speed in processing pathology results:

---

[122] ibid.

[123] ibid.

> During 1999, Corporal Summerscales processed 12,600 pathology tests. To date Corporal Summerscales has processed over 6,000 pathology tests for the year 2000. This is comparable to the workload for a Scientific Officer and three pathology technicians ...
>
> Corporal Summerscales has deployed twice with the Peace Monitoring Force in Bougainville ... On both of her deployments she performed admirably and was considered to have been an asset to the Combined Health Element.[124]

Kerry subsequently served in East Timor from 2001–2002 and in the Solomon Islands from July 2003 to January 2004. On her return she was commissioned as a Lieutenant and appointed a Scientific Officer.

In contrast to Bougainville, Kerry found her deployment to the Solomons more challenging. This time her working environment was a 'hospital afloat', with the Royal Australian Navy. As well as seasickness and the confined spaces she disliked the lack of freedom – if someone needed blood, she had to ask a nurse to unlock the blood fridge. On the rare occasions that she did go ashore, she again found that her gender was an asset in the cultural environment. She was able to provide healthcare advice to young women and deal with pregnancy complications.

In 2011, Kerry decided to retrain as a doctor and started a new phase in her medical career. Recently graduated from Flinders University, she is Captain Kerry Summerscales, RAAMC, posted to Edinburgh Defence precinct as a Medical Officer.

East Timor

In 1999, as East Timor gained its independence from Indonesia, the Australian Defence Force commenced an ambitious regional peace-enforcement operation; one that in extent and duration dwarfed all of Australia's previous peacekeeping efforts.

Located less than 600 km from the mainland of Australia, next to the Indonesian archipelago, East Timor comprises the eastern half of the island

---

[124] Citation sourced from the private records of Captain Kerry Summerscales.

of Timor. Division of the island has its roots in colonization. While the Dutch colonised the west of the island, the Portuguese colonised the east.

In 1975 Indonesia invaded. Portugal was emerging from its own revolution; and there was civil war between the supporters of political groups in East Timor. Fearful that a communist state could be installed on its doorstep, Indonesia declared East Timor to be its 27th province. It was a claim that the UN Security Council refused to accept – East Timor was regarded a non-self-governing territory under Portuguese administration.

On 30 August 1999, the East Timorese people voted overwhelmingly for independence from Indonesia in a referendum that was sponsored by the UN. Immediately afterwards, however, Indonesian militia launched a campaign of destruction, arson and murder that left 1,500 people dead and forced 300,000 refugees across the border into West Timor, controlled by Indonesia.

The UN responded with a resolution for a multinational force to restore peace and security and facilitate humanitarian assistance.[125] The force was initially called INTERFET (International Force for East Timor) and later UNTAET (United National Transitional Administration in East Timor). It was led by Australian Army General Peter Cosgrove.

As part its mission, the UN established a military hospital (UNMILHOSP) at Comoro airfield in East Timor in September 1999. The primary role of the hospital was to provide health support to the UN peacekeeping group, but was also authorised to provide emergency care to the local civilian population. The hospital served around 8,000 soldiers with the UN peacekeeping force and 4,000 civilian workers who were also with the UN contingent[126].

At the time, UNMILHOSP was the only fully functioning hospital

---

[125] Acting under Chapter VII of the Charter, the Security Council authorized the establishment of a multinational force under a unified command structure to restore peace and security in East Timor, to protect and support the United Nations Mission in East Timor (UNAMET) and to facilitate humanitarian assistance operations. Resolution 1264 (1999) was unanimously adopted and authorised the States participating in the multinational force to take all necessary measures to fulfil that mandate.

[126] Health Service Profile, United Nations Military Hospital, East Timor. Available at: http://www.defence.gov.au/health/infocentre/journals/ADFHJ_apr01/ADFHealthApr01_2_1_BC.pdf Accessed 16 June 2014.

on the island and demands on its 143 personnel were high. They came from a range of backgrounds and were multinational, including Egyptian, Singaporean and Australian personnel. Although staff numbers were relatively low, the hospital provided a level of service that was equivalent to a facility in a small Australian country town with two operating theatres, general and orthopaedic surgical teams, an intensive-care unit, two 25-bed general wards and an emergency room. The hospital also had a laboratory, radiology, pharmacy, outpatients department, physiotherapy, dental services and its own internal logistics support that included catering and laundry.[127]

The fourth rotation of Australians at UNMILHOSP was largely drawn from the 3rd Health Support Battalion (3HSB), with its HQ at Keswick Barracks in Adelaide. This unit has a long and proud history of supporting overseas deployments going back to 1916 and the 3rd Australian General Hospital, established at Turks Head, some 50 miles from Gallipoli. Since that time the unit changed role many times, but continued to provide support – from Abbeville on the Western Front in World War I to the Middle East campaign of World War II. In more recent times it has provided individuals and teams for deployments in Vietnam, Rwanda, Bougainville, East Timor, Iraq and Afghanistan.

The deployment of 3HSB to East Timor was significant both in military and medical terms. It was the first time that a Reserve unit was deployed on active service since World War II.

Colonel Vikija Andersons RFD

An officer in the Army Reserve, Colonel 'Vicki' Andersons assumed command of the UNMILHOSP on 23 February 2001. Her military service began after she completed medical school at the University of Adelaide. She was in her intern year when she was prompted to apply:

> We all received letters from the Director of Medical Services. He wrote to all the interns asking us if we were interested in joining the Army Reserve … I rang him up and said, 'Look, you've sent

---
[127] ibid.

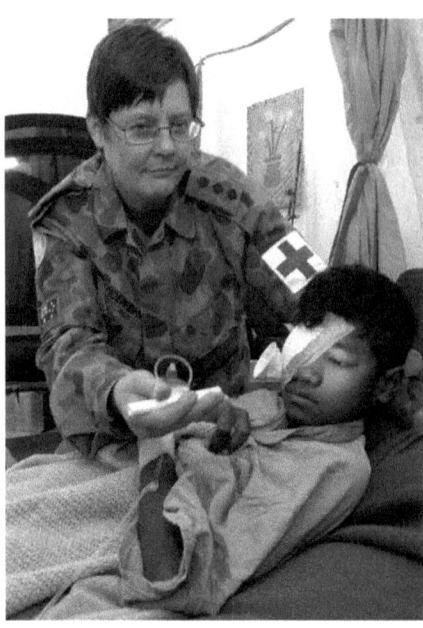

Colonel Vikija Andersons with 'Jonny' in East Timor. *Photo by Sergeant Bill Guthrie,* Army: The Soldiers' Newspaper, 13 September, 2001. *Image: Department of Defence.*

this letter to a girl'. And he replied; 'Come and see me anyway'. I did that and that's how it started. I signed up.[128]

From her early experiences as a Reservist Medical Officer, Colonel Andersons was surrounded by a wealth of clinical leadership from more senior specialists that had served during the Vietnam War. She progressed through the ranks of the Reserve, completing her officer courses. By the time of the East Timor intervention, Colonel Andersons was Commanding Officer of the 3rd Health Support Battalion (formerly 3rd Forward General Hospital).

Colonel Anderson's command in East Timor included the Egyptian and Singaporean contingents as well as South Korean guards. Such cultural diversity among the hospital staff provided challenges in communication: there was a need to coordinate care protocols and methods between the different national contingents. After hours, the multicultural environment

---

[128] Lipson N. Eye doctors with true ANZAC spirit. Mivision 26 March 2010 Available at: http://www.mivision.com.au/eye-doctors-with-true-anzac-spirit/ Accessed 4 November 2011.

also had its advantages – there were ample opportunities for cultural evenings and social functions.

The main illnesses of concern were malaria and dengue, both of which were highly prevalent during the wet season from November to March. There were also injuries – one involved a 15 year-old boy called 'Johnny' who Colonel Andersons remembers fondly. He suffered a serious penetrating eye injury after a fight with a friend. He had been treated by Australian medics at a UN border-post before being sent to Dili. Unfortunately his eye was unsalvageable and had to be removed, in order to prevent losing vision in his remaining eye:

> If you don't remove the affected eye then a patient can become totally blind. And that's what we had to do in Johnny's case.[129]

Colonel Anderson's distinguished career also included postings as Director of Health Services in South Australia and as the Senior Medical Officer for the Second Infantry Division based in Sydney and command of the 3rd Field Ambulance. In her civilian life, she became one of the few female ophthalmologists in South Australia.[130]

Major Suzanne Le Page Langlois

Joining the Army at aged 51 is not everyone's response to a 'mid-career crisis', but such was the choice of Suzanne Le Page Langlois.

Suzie joined the Australian Army Reserves in 1999. Aged 51, she was an unusual recruit – she had teenage children and an established career as both an academic and clinician. She was an Associate Professor at the University of Adelaide and Director of Radiology at the Royal Adelaide Hospital (RAH). Yet Suzie longed for more, wanting to test her skills in an even more challenging setting.

Suzie came from a military family reputedly descended from French nobility. She had an ancestor who was born illegitimate to a member of the

---

[129] Gilby, N., 'An eye for detail in Timor', *Army – The Soldiers' Newspaper*, 13 September 2001.
[130] Vikija Andersons was awarded an AM in the 2014 Australia Day Honours: For significant service to medicine as an ophthalmologist and surgeon.

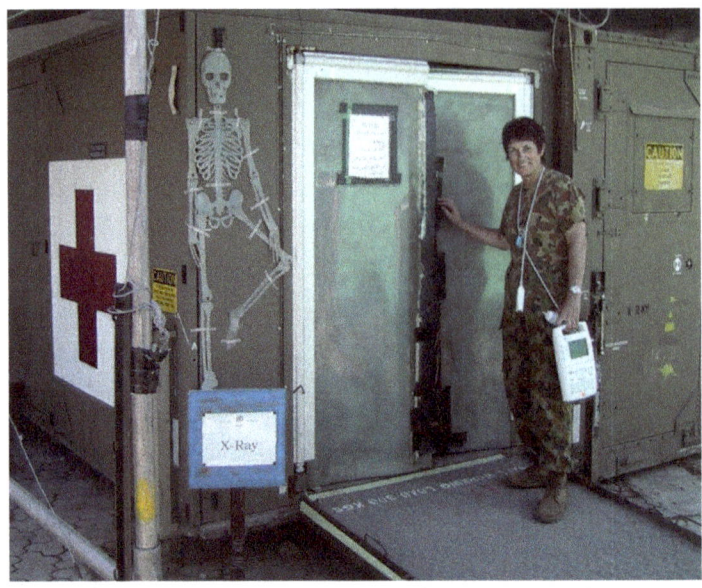

Major Suzanne Le P. Langlois in East Timor outside the UNMILHOSP radiology facility. She is holding the portable ultrasound used during her deployment. *Image courtesy of Suzanne Le P. Langlois*

court under Louis XVII. Such births were not uncommon and the child was brought up by the King's entourage and given the role of page-boy. This led to Suzie's unusual surname, and its acknowledgement of 'Le Page', often abbreviated to 'Le P'.

Both of Suzie's parents served with the Royal Australian Air Force during World War II. Her mother stayed on to assist with demobilisation – as a working mother, she became a role model for her daughter. Although Suzie was career-oriented, she did not consider the military until she was much older: "Going through school at Brighton High I remember looking out of the window at the Army Cadets. There was a never a question of joining them. It [just] wasn't something that girls could do," she recalls.[131]

It was during her work at the RAH that she became inspired by some of her colleagues and, she admits, envious of the work they were doing. Two colleagues in particular made her wonder if she could achieve more. Dr

---

[131] Extract from interview between Suzanne Le Page Langlois and Sharon Mascall-Dare recorded on 14 August 2013.

Other people's wars: peacekeeping near and far

Toby Thomas, a Colonel in the Army Reserve and former Commanding Officer of the 3rd General Hospital, was Head of the Intensive Care Unit. With him worked Dr Bill Griggs, a Wing Commander in the Air Force Reserve. Both had extensive deployment experience with the ADF. Bill was the director of retrievals at the hospital and would later play an important role in treating victims of the Bali bombings. Suzie looked on, inspired and determined:

> I was jealous of the men because they had been able to do military training in high school. I saw Toby Thomas, Bill Griggs and others at the Royal Adelaide and they would go off and fly out and pick up people and bring them back …
>
> We'd go up and look at the X-rays of their patients and I'd hear about their work. They were saving lives, they were achieving things and all I was doing was telling them what was wrong on the X-ray.[132]

Suzie considered all three services before deciding which to apply for. Eventually she settled on Army, put off by the underwater drills that she would have to pass to join the Navy.[133] Given her age, seniority and professional expertise in radiology, she was appointed as a Specialist Service Officer with the rank of Major. While her friends and family were incredulous at her decision, Suzie saw it as an obvious step: "I was looking for the next challenge. I had been involved in Neighbourhood Watch for many years and somehow thought that Army would be less demanding."[134]

Her naivety did not last. Suzie describes her training, at the Royal Military College Duntroon, as "brutal".[135] Confronted by an RSM shouting orders she did not understand and nights in a sleeping bag in sub-zero temperatures, Suzie's physical fitness and mental resilience were tested to the limit. More at home with stilettos, silver cutlery and fine china, Suzie had never slept in a sleeping bag until that freezing winter

---

[132] ibid.
[133] ibid.
[134] ibid.
[135] Ibid.

in Canberra. Although she was fit and regularly ran half marathons, as a middle-aged woman Suzie found it very hard to keep up with her fellow 20-year-old recruits or understand why she occasionally landed in trouble: "Not knowing the rules before you broke them was the part that upset me most," she recalls.

Undeterred, Suzie stuck it out and was asked to deploy to East Timor 18 months after her appointment. Given the choice of a six-week deployment as a radiologist specialist or a six-month tour as a radiographer-soldier, she chose the latter, seeing an opportunity to push her personal boundaries.

At first, the RAH were unwilling to lose her for six months. Even harder was leaving her family. As a single mother, she would be leaving her teenage children and her daughter, in particular, did not want her to go. She faced an agonising decision: leave her family and her job, or lose a once in a lifetime opportunity. It was hard, but Suzie persuaded her family and colleagues to let her go. With their blessing, she left Adelaide for pre-deployment training in Darwin in early 2001.

It was a different style of travel from what Suzie was used to. As a self-described 'five-star soldier', Suzie's previous life involved business-class airfares, French champagne, designer clothes and a wardrobe of shoes to rival the collection of Imelda Marcos. Swapping all this for an army uniform impregnated with insecticide instead of French perfume, and a rifle that she took everywhere, including to the shower, would be a challenge to most women. Suzie, however, shrugged it off: "One of the great benefits is that you don't have to wear make-up, contact lenses, high heels or stockings."[136]

As she packed her suitcase for East Timor, there was another reality check. Suzie did not just travel 'heavy', she travelled seriously heavy. The upper limit allowed for her luggage was 32 kg, "less than I usually take away for a weekend".[137] Armed with her suitcase, the counter staff recognised her immediately at Adelaide Airport – as a senior radiologist and academic of international standing she was one of their more frequent flyers. They had upgraded her automatically:

---

[136] Extract from interview between Suzy Langlois and Susan Neuhaus 24 February 2010.
[137] ibid.

All of us were in line and they said, "Oh, Dr. Langlois we've upgraded you to business class. You're in your usual seat 2A".

And I said, "Oh, you can't do that. I'm with the Army. I've got to go sit in the back", because the CO was sitting back in economy.

I do *not* fly economy. I mean, I do not fly economy. But, they had to put me back there. So back I went.[138]

While Suzie respected the protocols of hierarchy, she had, however, broken the rules with her luggage. Determined to use her portable ultrasound out in the field, Suzie had packed the $50,000 machine in the bottom of her trunk even though she did not have permission to do so.

In December 1999, Suzie obtained a new portable ultrasound machine, which arrived in Australia from the United States. The Royal Adelaide Hospital was, therefore, the first hospital in the country to use it for research and training purposes.[139] Developed for the US military, the machine was designed as a field device to be hand-held by soldiers or medics, producing high-quality, enhanced images that could be downloaded via satellite to physicians at base hospitals.[140] The machine stood to transform radiographers' diagnostic capability out in the field, but the ADF was not willing to insure Suzie's machine as non-issued (and expensive) equipment. Suzie knew that this was the ideal opportunity to pioneer the use of mobile ultrasound, but establishing this new technology in a conflict zone required determination and a degree of obstinacy in the face of opposition.

Within hours of her arrival, however, her decision to pack the machine paid off. On her first day at UNMILHOSP, Suzie was being shown the hospital's radiology equipment when a young man was brought in that afternoon after a road accident. With multiple injuries including two broken legs and a fractured skull, the man was sent to intensive-care,

---

[138] Extract from interview between Suzanne Le Page Langlois and Sharon. Mascall-Dare recorded on 14 August 2013.

[139] Le P. Langlois, Suzanne, 'Portable Ultrasound on Deployment', *ADF Health*, 2003 (4): 77–80.

[140] ibid.

but the doctor assessing him remained concerned. Aware that Suzie had 'smuggled' a portable ultrasound into the hospital, the doctor asked her to fetch it from storage. Within minutes they knew what was wrong: there were large volumes of blood in the man's abdomen from a ruptured liver and spleen.

> We went down to the Q Store and got my trunk, because nothing had been unpacked ... and it [the ultrasound machine] just had enough charge to do an abdominal ultrasound and he was in deep trouble. He went straight to theatre. It saved his life.[141]

From then on, Suzie's ultrasound became an essential piece of equipment at the hospital. The treating physicians were impressed and there were no further questions about where the machine had come from. It proved invaluable during resuscitation and for assessing the injuries of UN staff and soldiers.[142] Although it was outside of the UN's mandate to treat non-emergency cases among local civilians, the machine was taken to Dili General Hospital once a week to train local medical staff – during each session around 25 local people would be examined and, where necessary, referred for treatment.[143]

The opportunity to diagnose and treat conditions that she would never encounter in Australia was a highlight of Suzie's deployment: "I didn't want to leave ... for the first time I was practising real medicine and making a difference. I saw diseases I would never have seen at home – tuberculosis (TB), dengue, heart conditions and obstetric challenges, where what I was doing was really worthwhile." She also treated Xanana Gusmao, and was impressed by his charisma. "I could see why people followed him", she says.[144]

As a consultant radiologist, Suzie would normally have had a radiographer to manage the technical aspects of developing X-rays,

---

[141] Extract from interview between Suzanne Le Page Langlois and Sharon Mascall-Dare recorded on 14 August 2013.
[142] 'Portable Ultrasound on Deployment', *ADF Health*, 2003.
[143] ibid.
[144] Extract from interview between Suzanne Le Page Langlois and Sharon Mascall-Dare recorded on 14 August 2013.

while she reported them. In East Timor Suzie had to do both; a role that challenged her technical skills, but saved films being sent back to Australia for reporting. Her employment classification however, was not without its problems and issues of pay and entitlement took some time to resolve.

One of her most memorable cases involved an elderly man, who had been diagnosed with tuberculosis (TB). Since he was thought to be infectious, he had been isolated from his family, unable to go home, confined to a bed in a TB ward for six months.

> I examined his tummy and he had lymph nodes around his aorta and, especially down in the lower pelvis, they had this funny whorled appearance, which is typical of testicular carcinoma. I said, "I don't think he's got TB – I know he hasn't got TB. I think he's got metastatic testicular carcinoma. Has anybody examined his testes?"
>
> They ushered me out of the room. They pulled a curtain around and they examined his testes. And he had this huge lump in his groin. He had testicular carcinoma. Now, I'd given him a death sentence … He wasn't going to get better, but it meant that he could leave the ward and go home to his family to die.
>
> That was probably one of the most useful things I did. But all of it was worthwhile.[145]

Suzie also set up a monthly 'Women Doctor's Network' in East Timor to encourage collaboration between her colleagues and local practitioners. Personally, she developed a warm and lasting friendship with Gigi Lima, one of few women in the Portuguese Army and a plastic surgeon.[146]

She was also known for maintaining her individuality during her deployment:

> The whole time that I was there I kept my sense of self … in my

---

[145] ibid.

[146] It was only in the late 1980s that a few women with special qualifications, such as doctors, lawyers, and engineers, were taken into the Portuguese officer corps. As of 1991, fewer than 100 women served in the Portuguese armed services, fewer than any other country of NATO. See: http://www.globalsecurity.org/military/world/europe/pt-personnel.htm Accessed 16 June 2014.

pack, there were pink sheets to go in my sleeping bag, a pink frilly pillow case to go on my pillow, a plastic wine glass so that I could have my Diet Coke in a wine glass. I took my own cutlery because I do not eat plastic and I do not do plastic.

When I give lectures about it ... they love to hear about how I coped: being an older person and still being me.[147]

On her return to Australia, however, Suzie realised that the experience had changed her: "It was not so much a case of PTSD, more one of a changed perspective on medicine and work – it was just so boring when I came home."[148] The RAH greeted her with streamers and balloons, but Suzie saw the limitations of her previous work – she had a heightened sense of purpose and knew she could achieve more. She moved to Townsville and took up a new role, establishing a new radiology department to service northern Queensland. She also continued to educate others in the use of portable ultrasound, overcoming opposition from radiologists who were not keen for other clinicians to take the place of radiologists in using the new technology.

Today, Suzie has semi-retired from medicine. Her lakeside property in Adelaide pays homage to her military service in photographs and memorabilia. It also has four kitchens to support her new hobby. In retirement, Suzie has retrained as a chef, and regularly throws dinner parties for her friends. She now uses her eye for the visual – once reserved for ultrasound analysis – to create works of artistic gastronomy. Her regular dinner guests include the likes of Gabriel Gaté – a far cry from lining up for meals on paper plates with plastic cutlery in the mess hall in Dili.

Lieutenant Joanne Marks

While Suzie's decision to join the military in her early 50s raised eyebrows, Jo Marks encountered a similar reaction at an earlier stage in her career.

Jo also deployed to East Timor as a radiographer in 2001, replacing

---

[147] Extract from interview between Suzanne Le Page Langlois and Sharon Mascall-Dare recorded on 14 August 2013.

[148] Extract from interview between Suzie Langlois and Susan Neuhaus 24 February 2010.

Other people's wars: peacekeeping near and far

Captain Joanne Marks in the Solomon Islands while on Operation Anode, 2003.
Image courtesy of Joanne Marks

Major Suzie Langlois while she was on leave. Jo, too, came from a military family. Her father served with the Royal Australian Corps of Signals and she had spent much of her childhood overseas.

After her father's posting to Singapore, where Jo was born, and Hong Kong, Jo's family returned to Australia and settled in Queensland. Although her parents expected her to become a nurse on leaving school, she decided to do something different following an offer from her uncle, who ran a film production company.

For the next five years Jo worked on television commercials, documentary films and corporate launches. One of her most exciting projects was working on a television commercial featuring Tina Turner to promote the Australian Rugby League. "For a young person I had a lot of responsibilities," she recalls. "I had to organise equipment, logistics and I had a bent towards it. I did it well."[149]

Despite her aptitude for the entertainment business, however, Jo felt that she did not truly belong. Her upbringing had made an impact and one day, without telling anyone, she decided to join the Army. "I knew that everyone would say, 'What are you thinking?' But the film industry weren't my people," she explains. "In a way, my dad raised us like soldiers

---
[149] Extract from interview with Jo Marks by Sharon Mascall-Dare on 1 December 2013.

– we were very focused and he was a wonderful man. I felt like I belonged in a military environment."[150]

When Jo joined the Army in 1990, the response from her fellow recruits at Kapooka was similar to that at home: genuine surprise that she had left the film industry to become a soldier. Her platoon commander tried to push her into a public relations role with the Army, but Jo resisted. "That's where I'd come from, that's not where I wanted to go," she says. "I wanted to go to RMC [the Royal Military College]. That's all I wanted to do."[151]

Jo thrived at Kapooka and was named 'Most Outstanding Recruit'. Although she found the course arduous, challenging and physically tough, she believes her upbringing helped. As a child her father had routinely pushed her to achieve more and challenge her personal boundaries. At the end of her training she chose the RAAMC, and completed her medical assistant training at Portsea. Again, she excelled and was awarded 'Student of Merit'. The Army was clearly the "right fit".

Over the next few years she completed studies in radiography under the Army's long-term schooling scheme and was commissioned as an officer with the rank of Lieutenant. Shortly after finishing her studies she was offered the opportunity to deploy in East Timor.

From the moment she arrived Jo was aware she was entering an area of operations. On the journey from the airport to get her identity card:

> There was a staff sergeant in charge of the minivan, we're driving through the streets and there were burning buildings all around. It looked like a warzone[152].

From the first day of her deployment, Jo learned to deal with whatever "came through the door". Together with her assistant, an Egyptian sergeant, called Sharif, she dealt with the steady stream of UN staff and local people needing X-rays, working closely with reservists and medical specialists.

The ability to provide diagnostic images within seconds is crucial to saving lives in high-intensity trauma situations. In field environments,

---

[150] ibid.
[151] ibid.
[152] ibid.

Other people's wars: peacekeeping near and far

X-ray machines are mobile, and can be used in the resuscitation bays to identify chest injuries, fractures and foreign objects, such as bullets. Jo used mobile apparatus as well as fixed equipment housed in a Brunswick shelter (shipping container) at the UNMILHOSP in Dili. The processing of images was carried out via wet processing, using benchtop film processors. Deterioration of the film and chemicals was an ongoing challenge, especially in the tropics with high humidity and large temperature variations between the middle of the night and the middle of the day. It required skill to consistently produce diagnostic images in such conditions. Maintaining radiation safety was also a challenge. Jo used a processor located within a light-tight darkroom contained within the X-ray shelter.

On the few occasions that she ventured outside the hospital, she was struck by the destruction of the landscape and the commitment by the East Timorese people to pursue their way of life despite chaos and violence. She remembers going for a walk one Sunday morning and seeing local people attending church in pristine white clothes, emerging from the tiny tin huts that were their homes. "Regardless of their environment, they were still focused on maintaining high standards and beliefs. They got on with their lives," she recalls.

Jo was also struck by the proximity of the conflict to Australia. On her last night at the UNMILHOSP before flying home, she was called in to assist with the case of a young Timorese girl who had been hit by a military vehicle. The girl had sustained multiple, serious fractures, particularly to her pelvis, and Jo worked through the night to provide the X-rays required. Suddenly, her work was done and she stepped outside, it was daylight and she had only minutes to get to the airport. Just a few hours later she found herself in Darwin airport:

> It was culture shock of the highest order. The next thing I knew I was in Darwin. My head was spinning, but the girl's name has always stayed with me.
>
> I can still see the image of her hips on the X-ray. And I can still hear the crying of her family outside the gates of the UN [compound].[153]

---

[153] ibid.

Solomon Islands

Just after dawn on 24 July 2003, the first Hercules touched down in Honiara with lead elements of the Regional Assistance Mission to Solomon Islands (RAMSI). With 1,400 troops, 300 police and officials from the nine Pacific Forum countries, the role of the mission was to bring peace after four years of increasing violence. Ethnic, social and criminal disorder in the Solomon Islands had cost 200 lives and led to the near collapse of the national government and economy. Some 20,000 people were displaced – forced to flee their homes.[154]

Earlier that year, in April 2003, the rule of law had all but collapsed. The then Prime Minister, Sir Allan Kemakeza, sought foreign assistance; he had no other choice. The roots of the conflict lay in long-standing tensions between ethnic groups, exacerbated by the arrival of economic migrants from the island of Malaita to the main island, Guadalcanal. As tensions and violence spread, authorities lost control. Australia consulted its Pacific neighbours to provide a solution.

Its response was RAMSI, comprising personnel from 15 Pacific nations, led by the Australian Government. Civilian police were a key component, backed by a military contingent, including infantry from Australia, New Zealand and Fiji as well as Tonga and Papua New Guinea.

It was a unique mission for Australia, drawing on the lessons learned from Bougainville and other peacekeeping missions in the post-Vietnam era. It was a 'whole of government' response with an alternative model – it was not a military operation, but a civilian intervention backed by military security. It had its critics, but concerns were soon quelled. It was a mission committed to the restoration of law and order in a nation crippled by conflict.

At the same time as the Hercules landed, the amphibious ship HMAS *Manoora* loomed off the coast. *Manoora*'s heavy presence signalled that support was near. Within the year, warring militias would be disarmed; economic stability, effective governance and security would also return.

---

[154] See: Claxton, Karl, 'RAMSI tenth anniversary: thinking about intervention and amphibs', Australian Strategi Policy Institute. Available at: http://www.aspistrategist.org.au/ramsi-tenth-anniversary-thinking-about-intervention-and-amphibs/ Accessed 30 May 2014.

Other people's wars: peacekeeping near and far

Captain Joanne Marks and Sergeant Kerry Summerscales CSM

Although she was raising her eight-year-old son on her own, and relied on family and friends to care for son in her absence, Jo Marks was keen to build on the skills she had learned in East Timor. When she was offered the opportunity to go to the Solomon Islands for nine-month rotation from 2003–4, she accepted:

> We didn't really get much preparation. We knew we were on notice. You're always prepared to go. Then we were told definitely, we're leaving tomorrow. That was 2 o'clock in the afternoon. I rang my mother … I picked my son up from school. I took him round to my brother's and we had dinner. My little fellow fell asleep on the lounge and I was gone. I was gone for nine months.
>
> When I left the Army, I realised I was not the most nurturing mother … I always tried to do my best. But he [Jack] is so proud. He has no negative words whatsoever. He did suffer when I was in the Solomons. He woke up the next morning and I'd gone. The kids at school knew I'd been deployed and they would say, "Your mum will get killed over there."
>
> After around eight months, I came home on leave. I caught a taxi and I could see my parents' property. I could see Jack out in the paddock playing with a ball. I said to the man driving the car, "Stop here," and I got out, climbed over the fence and walked across the paddock.
>
> Jack came over and said: "Are you my mummy?" That's when I realised the toll it had taken on him, with me being away.[155]

Jo's experiences resonate historically and in a contemporary context. Mary Thornton and Joan Refshauge both made the decision to send their sons to boarding school during the war and post-war service in the 1940s; many women since have left their children to deploy on operations with the ADF.[156] Given the increase in the numbers of women, including

---

[155] Extract from interview with Jo Marks by Sharon Mascall-Dare on 1 December 2013.

[156] See chapters 3 and 4 for Major Mary Thornton and Captain Joan Refshauge; Susan Neuhaus's experiences are recounted in the Epilogue.

mothers, who have deployed on overseas operations in recent years, and are likely to deploy in future, Jo's mixed feelings are prescient.

Life on the ship was hard. There was occasional tension between Navy and Army personnel as the months went by and there was little opportunity to disembark. Jo served alongside Kerry Summerscales, who had also returned from East Timor. Kerry found the conditions on ship particularly hard:

> I hated it. I hated it because I'm not used to confined space … it's a different culture from what I'm used to.[157]

The majority of Jo and Kerry's duties were undertaken on board HMAS *Manoora*, moored off the coast of Honiara. Suffering from repeated bouts of seasickness, the women struggled to live and work in the confined space – they longed for the respite of dry land during short field visits on shore. When they did disembark, interaction with local people was a highlight. Kerry recalls one incident in particular:

> One day I was on shore washing my cams and a local woman came up to me. She said there was a woman in a nearby village who had given birth to a baby girl. Then she said, "Basket bilong pikinini still inside." And I thought: "The basket belonging to the baby is still inside. Oh, good God! That's a retained placenta!"[158]

Fortunately, the woman's village was only half an hour's walk away. Together with a male medic, Kerry found the woman and put in a drip, seeing that she was haemorrhaging badly. Knowing not to pull on the umbilical cord, she gently massaged the woman's fundus and the placenta came away. She then arranged medical evacuation to a nearby hospital.

As the months went by, Jo and Kerry relied on each other increasingly for support and friendship. On the rare occasions when they took a walk through the local town they were reminded of the reasons they were there, despite the difficulties they encountered.

---

[157] Extract from interview by Sharon Masacll-Dare with Kerry Summerscales on 21 October 2013.
[158] ibid.

> One time, we were walking through town and we're bleating away. 'Whinge, whinge, whinge,' as you do.
>
> This woman stopped us and she said, "Me must thank you".
>
> I said, "What for?"
>
> And she said, "My daughter and I, we can walk down the street now, not fear rape because you're here. So, me thank you."
>
> Both of us left dumbfounded. As she walked off we both looked at each other and said, "Well, that put our complaints into perspective".[159]

After nine months on deployment to the Solomon Islands, Jo and Kerry returned to Australia early in 2004. RAMSI was to continue for another nine years, concluding in 2013.

While Kerry continued her military career, Jo was medically discharged in 2007. Despite a number of operations and rehabilitation, Jo was unable to fully recover her fitness following a training accident in 2002, after her deployment to East Timor. Reluctantly, she moved on and took up a new career as a radiographer and sonographer in the civilian health sector. Now grown up, her son, Jack, has become one of her closest friends – he reminds her, regularly, that he is proud of her 17-year military career.[160]

For Kerry, her lasting memory of the Solomons is the lights on shore that were visible from the ship. "When we arrived there were hardly any lights," she recalls. "But at the end the local people had light in their homes, they had some form of electricity". After months of darkness, peace was being restored.

---

[159] ibid.

[160] Extract from interview with Jo Marks by Sharon Mascall-Dare on 1 December 2013.

CHAPTER SEVEN

A new kind of warfare: the rise of global terrorism

Every war is the same. It kills people; it changes people's lives; it destroys people's villages ... war is just destructive. It doesn't achieve anything, from my point of view, but the devastation stays forever and ever.

Dr Tam Tran, a former refugee from Vietnam, on her service as an RAAMC Medical Officer in Iraq.[1]

Once they know that you're good at your job then you're fine, no matter female or male ... It was more, "can you do the job?", "do we want you as our medic?", "can we trust you?".

Corporal Jacqui de Gelder, on her experiences in Afghanistan as part of Operation Slipper in 2009–2010.[2]

KURDISTAN

Captain Thi Thanh Tam Tran CSM

Captain Tran watched the stream of people flow down the mountain. It was hot, extraordinarily hot, and the thermometer was close to hitting 50 degrees. It was a dry heat, there was little humidity, and on occasion its intensity caught Tam unaware. Heat stress was always a danger during this deployment; it was part of the job.

---

[1] Tran, Tam, extract from an interview recorded with Sharon Mascall-Dare on 15 August 2013.

[2] Extract from interview with Jacqui de Gelder by Sharon Mascall-Dare recorded on 17 December 2013.

A new kind of warfare: the rise of global terrorism

The mountain was close to the border between Iraq and Turkey. From May to June 1991, 75 Australians, including Captain Tam Tran and fellow medical officer Captain Jane Morris, were deployed on its slopes as part of a humanitarian mission called Operation Habitat. The first Gulf War had ended in February that year, and Iraqi forces had withdrawn from Kuwait. In the aftermath of the conflict, armed Kurdish resistance forces had tried to seize control in northern Iraq, only to be brutally crushed by Saddam Hussein's military.

Half a million Kurds fled north, into the mountains on the border with Turkey; a million fled east to Iran. Previously, in March 1988 – the last days of the Iran–Iraq War – the Iraqi Government had unleashed a poisonous arsenal of chemical weapons against the Kurdish people, killing 5,000 people and injuring 10,000 more.[3] That attack left a legacy of medical complications, diseases and birth defects; it was an act of genocide that the Kurds feared could be repeated. There were reports that Iraqi loyalists were again using chemical weapons against rebel forces. Once the Kurdish uprising was crushed, the exodus of refugees began in earnest, with tens of thousands seeking refuge in the remote and inhospitable terrain of the mountains.

The Australian Defence Force contingent was located at Gir-i-Pit, around 30 kilometres north of Dahak in northern Iraq. Nestled in a valley and surrounded by mountains up to 3,000m high, the area is snow-covered in winter. By May, when the Australians arrived, temperatures were already sweltering, in the high 40s to 50 degrees C. Accommodation was in tents defended by weapon fighting pits and barbed wire.[4] As well as the threat of

---

[3] Hiltermann, Joost, *A Poisonous Affair: America, Iraq and the Gassing of Halabja*, Cambridge University Press, Cambridge, 2007. A combination of chemical weapons were used, including mustard gas and nerve agents. See Gosden, Christine and Gardener, Derek, 'Weapons of Mass Destructions – Threats and Responses', *BMJ*, 2005 August 13, vol. 331(7513), pp. 397–400.

[4] 'Case For the awarding of the Australian Active Service Medal 1975 Clasp IRAQ to those ADF members who served with the Australian Contingent to Northern Iraq on OPERATION HABITAT (NATO OPERATION PROVIDE COMFORT)', submission to the Government Review into the Veteran Entitlement Review Committee (Clarke Review) by the Australian Peacekeeper and Peacemaker Veterans' Association, Annex A to Part F, p. 4. Available at: http://www.peacekeepers.asn.au/veterans/submissions/OP%20HABITAT%20AASM%203%20Jan%2010.pdf Accessed 10 January 2014.

attack from Scud missiles[5] there were unexploded chemical weapons.[6] The heat too, was a persistent and formidable enemy.

Captain Tam Tran was one of five women deployed to the region with the Australian medical contingent.[7] There were four medical teams, each with five personnel, including a medical officer, nursing officer and three medical assistants – one was an NCO.[8] Each team also had an interpreter attached. There was limited information available to the teams before their departure about what to expect. The humanitarian crisis in the mountains was changing rapidly. The teams relied on information from military medical and health intelligence sources as well as civilian aid agencies. They also spoke to medical staff who had previously worked with refugees.[9]

Tam was a medical officer. She had joined the Army during her medical school years at university in Queensland. She was one of six children and her parents were unable to fully fund her medical degree. "They were both factory workers and they found it really hard," said Tam. "I looked around for financial support and the Defence Force gave the best support in the scholarship system, so I took that up."[10]

While Tam's decision to join the ADF was primarily pragmatic, it was not without irony. Tam grew up near Saigon, and she remembers being

---

[5] The Scud is a of long-range surface-to-surface guided missile that can be fired from a mobile launcher. The word 'Scud' is used for a series of missiles originally developed in the USSR – the word was a codename for those missiles assigned by NATO. The Scud developed a fearsome reputation due to its accuracy and destructive power.

[6] Submission to the Government Review into the Veteran Entitlement Review Committee (Clarke Review) by the Australian Peacekeeper and Peacemaker Veterans' Association, 11 March 2009, pp. 29–30. Available at: http://www.peacekeepers.asn.au/veterans/submissions/APPVA%20Clarke%20Review%202009%20Main%2011%20Mar%2009.pdf Accessed 10 January 2014.

[7] Tam served alongside Captains Jane Morris, also a Medical Officer, and Karen Bayliss, a Nursing Officer. Sergeant Sandra White and Lance Corporal Julia Thomas were also part of the medical contingent.

[8] Little, Mark and Hodge, Jonathon, 'Operation Habitat, Humanitarian aid to the Kurdish refugees in northern Iraq', *Medical Journal of Australia* 1991; 155:2 (16): 807–812.

[9] Tran, Tam, extract from an interview recorded with Sharon Mascall-Dare on 15 August 2013.

[10] ibid.

A new kind of warfare: the rise of global terrorism

treated by Australian aid teams during the Vietnam War.[11] Tam's father had served with the South Vietnamese Army and the family fled Vietnam as refugees in 1976, a year after the fall of Saigon. Tam was about "nine or 10 years old" at the time and her memories of those years are, by her own admission, traumatic: "I grew up at the height of the Vietnam War and saw the devastation ... I was one of those boat people who came all the way to Australia."[12] She recalls the 56-day open sea voyage as being something of a childhood adventure, "but a harrowing and dangerous time for my parents and everyone else".[13]

Now, as she watched the thousands of Kurdish refugees descend the mountain in northern Iraq, Tam was reminded of her own experience, standing outside her parents' grocery shop on the outskirts of Saigon.

> As a child, I remember, just after the Vietnam War, [it was] about March or April 1975, when the south fell to the north. There were streams of people coming – walking along the road – because we lived close to the main road. We had a house, and mum had a little grocery shop ...
>
> People were just streaming down the road because they were trying to get to Saigon, and we were on the main highway that goes there, so there were lots of cars, lots of people walking, and they would come by to get water and food.
>
> As a little kid, you know, you just looked at them and thought "Wow." It's like a procession, like a parade. At the time I didn't realise what was happening or how devastating it was. People were leaving their belongings behind and were just trying to get south. And when we went to Iraq I was seeing similar things. People were streaming by and just walking, getting back from the mountain to their villages.
>
> I realised what was going on ... and that really, really brought it back to me about the war. And, you know, every war is the

---

[11] Little and Hodge.
[12] Tran, Tam, extract from an interview recorded with Sharon Mascall-Dare on 15 August 2013.
[13] Private correspondence between Susan Neuhaus and Tam Tran dated 17 May 2010.

same. It kills people; it changes people's lives; it destroys people's villages.

That really came back to me.[14]

At night, when the heat of the day had died down, Tam's memories of her childhood would again surface. During the day she kept them at bay: she was focused on the job. Looking back, with hindsight, she believes her past helped her to empathise with the Kurdish children, in particular. They were innocent, smiling and playful: kids who had nothing of their own, from abandoned villages, destroyed by the fighting. "Those are the pictures that stay with you," she recalls.[15]

Each morning, Tam's team would take to the rough, cobbled roads – with occasional respites of tarmac – seeking a base for their temporary medical clinic. Rising early, the team loaded up their Land Rovers with medical supplies, equipment and stores and would scout for locations: stopping multiple times on route, setting up under a tree or in a building, wherever there was greatest need.

While one team remained on base, the other three travelled up to 200 kilometres every day treating 60–100 patients – some 3,000 people received care from the Australians during the month and a half long operation. The refugees came from a range of backgrounds – university lecturers were subjected to the harsh, arid conditions alongside subsistence farmers.[16] The majority of patients were children.[17] When they encountered a 'village' there was often nothing left: buildings and infrastructure were in ruins and tents housed entire families. Negotiations with local elders to secure locations were not always straightforward. Although Tam was in charge

---

[14] Tran, Tam, extract from an interview recorded with Sharon Mascall-Dare on 15 August 2013.
[15] ibid.
[16] Little and Hodge.
[17] 'Case For the awarding of the Australian Active Service Medal 1975 Clasp IRAQ to those ADF members who served with the Australian Contingent to Northern Iraq on OPERATION HABITAT, (NATO OPERATION PROVIDE COMFORT)', submission to the Government Review into the Veteran Entitlement Review Committee (Clarke Review) by the Australian Peacekeeper and Peacemaker Veterans' Association, Annex A to Part F, p. 4. Available at: http://www.peacekeepers.asn.au/veterans/submissions/OP%20HABITAT%20AASM%203%20Jan%2010.pdf Accessed 10 January 2014.

of her team, the elders usually refused to talk to her since she was female. "Luckily I was assigned a nursing officer who was male. He would have to go and negotiate and explain, through an interpreter, that although he was negotiating the ultimate decision was mine. After a bit of to- and fro-ing, they eventually understood."[18]

Shade was always a priority and the wind provided welcome relief, when it came. Many of the conditions Tam's team encountered related to the heat and the lack of sanitation. Heat stroke, typhoid and malaria were common. They provided sanitised drinking water supplies for villagers, but much of their time was spent educating local women about nutritional requirements and the need to sterilize baby bottles.[19]

The operation was classified as "hazardous"[20] with some fifteen million landmines and unexploded ammunition scattered throughout the Kurdish area of operations and a constant threat of biological or chemical attack. At times the teams encountered hostile Iraqi forces; there were also Kurdish para-military with machine guns and field artillery.[21] They often encountered gunshot wounds that had been left untreated and had become infected. Some of those armed with AK-47s were children.

Although she rarely felt in danger herself, in the back of her mind the risks were always there: "We were armed and, obviously, we had our rifles with us all the time ... the immediate danger we always had to look out for was stepping on landmines. We were always wary of staying on the designated path and not wandering off. That was always a big danger."[22]

Relations with the British were strong. Landmine injuries were referred to the British, who operated a field hospital nearby. The hospital was only 20–30 minutes' drive from the Australians' base camp and Tam would often drop by to discuss individual cases and catch up on local news.

---

[18] Tran, Tam, extract from an interview recorded with Sharon Mascall-Dare on 15 August 2013.

[19] Williams, Lesley, *No Better Profession, Medical Women in Queensland 1891–1999*, Williams 2006. p. 163 and private correspondence between Susan Neuhaus and Tam Tran dated 17 May 2010.

[20] Williams p. 163.

[21] Clarke Review. pp. 29–30.

[22] Tran, Tam, extract from an interview recorded with Sharon Mascall-Dare on 15 August 2013.

Captain Thi Thanh Tam Tran in Northern Iraq during Operation Habitat in 1991.
Image courtesy of the Australian War Memorial, CANA/91/010/09.

The British had specialists in general surgery, anaesthesia, paediatrics and infectious diseases; there was a spirit of cooperation with weekly meetings of the 'Northern Iraq Medical Society' to compare notes and case studies. Tam describes the primary and preventative care provided by her team as "GP" (general practice) level; at times they also acted as an "ambulance", taking more serious cases to hospital. Operation Haven – the British commitment – and Operation Provide Comfort – the United States contribution – formed the backbone of an international coalition effort, one that included Spain, Canada and Turkey as well as Australia.

Eventually, as Tam became known to some of the local elders, they included her in their discussions. In a break with tradition, some allowed her to drink tea with them, a ritual and privilege usually reserved for men alone. Throughout her deployment however, Tam remained concerned about the treatment of local women. When villagers arrived for treatment, the men and children would line up while the women would hold back. "It was not until we packed up that they would come and ask for this or that ... we realised that women were not getting their fair share of treatment. So, in subsequent clinics we would negotiate a time dedicated to just treating female patients. That way the females were treated and looked after like everyone else."[23]

As the weeks went by, Tam and her team could see that they were making a difference. The Kurdish people were returning to their homes, shops were re-opening and children were returning to school. The Australians had established a rapport with local people and their makeshift clinic had become a reliable, if temporary, haven. It was a place that families could turn to, day after day. Then, without warning, they had to leave. For security reasons, they had to leave quietly as there was concern that the local population would react badly if they knew:

> I was a bit devastated. I thought that was a bit harsh for everybody. We didn't get a chance to say goodbye to the people, and they didn't get a chance to say goodbye to us. It was quite harsh. But then, you just did what you had to do. Obviously, someone else knew the bigger picture, and we – obviously – had to stick to that.[24]

In 1992, after her return to Australia, Tam was awarded the Conspicuous Service Medal for her service in Iraq, in recognition of her "meritorious achievement or dedication to duty in a non war-like situation".[25] The medal had been established only three years previously, in 1989, and Tam was one of three women to receive the honour that year. Australia's Vietnamese community considered her a heroine. Her story was one of triumph over adversity – as a 'boat person' she had evaded pirates near Thailand; as a migrant she had experienced racism at school.[26] She was asked to give interviews and speeches to community groups; she was

---

[23] ibid.

[24] ibid.

[25] See http://www.itsanhonour.gov.au/honours/awards/medals/conspicuous_service_medal.cfm Accessed 14 June 2014.

[26] Tam's experiences as a refugee and migrant to Australia have been used as a teaching tool in Australian schools. See http://www.quia.com/files/quia/users/sueangelo/year10/im_postwar.pdf Accessed 10 January 2014.

considered a role model to young Vietnamese women. Tam's reaction remained self-effacing, however:

> There were people there, people to be treated, and you just did it ... every speech that I gave, [I said] I didn't think I did anything different in Iraq: I just went and did what I was asked to do.
>
> Now that I'm older, wiser, and have seen even my own daughter growing up, the achievement – when you look back – was quite tremendous, even though I didn't think it was at the time.[27]

After her return to Australia, the Army sent Tam to Bangkok to complete a Post-graduate Diploma in Tropical Medicine and she was later posted to Lavarack Army Barracks in Townsville.[28] She left the ADF at the rank of Major and now works as a general practitioner in a family practice in a Brisbane suburb. While she finds "GP Land" occasionally mundane, she believes that her memories and experiences have given her a unique sense of empathy for her patients:

> I think it enriched my life. I think my service in Iraq helped me to understand and have faith in human compassion and the bigger picture of mankind.
>
> Despite the devastation of war, there is a capacity to help other human beings and to connect on a normal human level.[29]

BALI

Lieutenant Colonel Su Winter CSC

On the morning of 12 October 2002, Su Winter was watching CNN over breakfast in Darwin. Suddenly, the news came in that a bomb had exploded during the night in Bali.

The year before, on 11 September 2001, Usama Bin Laden had

---

[27] Tran, Tam, extract from an interview recorded with Sharon Mascall-Dare on 15 August 2013.
[28] Williams, p. 163.
[29] Tran, Tam, extract from an interview recorded with Sharon Mascall-Dare on 15 August 2013.

masterminded the worst terrorist attack on US soil, killing 2,973 people in coordinated hijackings that destroyed the World Trade Centre's Twin Towers in New York and attacked the Pentagon in Washington, DC.[30] When President George W. Bush declared a "war on terror" in response, Australia invoked the ANZUS Treaty and Australian military personnel joined 'the coalition of the willing'.[31] It was a decision that would lead to long-term involvement in the Middle East Area of Operations (or MEAO); it also made Australia a target. Along with her coalition partners, Australia was now an enemy of Usama Bin Laden's al-Qaeda terrorist network, including Jemaah Islamiyah, a militant Islamist terrorist organisation operating in Indonesia.[32]

Just after 11pm a suicide bomber detonated a backpack bomb inside a crowded nightclub, Paddy's Pub in Kuta, a popular destination for Australian tourists. As terrified patrons fled into the street, a second and much more powerful car bomb exploded 10 to 15 seconds later outside the Sari Club, on the other side of the road.[33] There was carnage: 202 people died as a result of the attacks, including 88 Australians.[34] More than 200 people were injured, many with severe burns, blast and shrapnel injuries. Both clubs were destroyed in the attacks along with neighbouring buildings; there were shattered windows several blocks away. The car-bomb explosion itself left a deep crater.

The local Sanglah Hospital was ill-equipped to deal with the scale of the disaster and the number of casualties, particularly burn victims. As

---

[30] National Commission on Terrorist Attacks upon the United States, *The 9/11 Commission Report*, Norton and Company, New York, 2011, p. 432.

[31] Australian War Memorial, 'Afghanistan, 2001–the present'. Available at: http://www.awm.gov.au/atwar/afghanistan/ Accessed 14 January 2014.

[32] For more information about the foundations and operations of Jemaah Islamiyah see: http://www.nationalsecurity.gov.au/agd/WWW/nationalsecurity.nsf/Page/What_Governments_are_doing_Listing_of_Terrorism_Organisations_Jemahh_Islamiyah Accessed 15 January 2014

[33] See: 'Operation Alliance', Australian Federal Police. Available at: http://web.archive.org/web/20060618005353/http://afp.gov.au/international/operations/previous_operations/bali_bombings_2002 Accessed 15 January 2014.

[34] See: *ibid* for casualty figures; for details of injuries see Cook, Steven, Smart, Tracy and Stephenson, Jeff, 'Learning the hard way: Australian Defence Force health responses to terrorist attacks in Bali, 2002 and 2005', *ADF Health*, 2006; 7(2): 51–55.

local clinics were overwhelmed, volunteer Australians and bystanders did what they could to assist, but the scene was one of complete chaos. So many people had been injured that some were placed in hotel swimming pools to ease the pain of their burns. Others found themselves sprayed or hosed down with water – water not fit for drinking, far less for pouring on raw burns.

Within minutes of the explosions, an off-duty Australian Army officer used his mobile phone to notify Army duty personnel in Australia. Meanwhile, as the casualties piled up, Australian diplomats went into action.

The Prime Minster, John Howard, was notified at 5am and a crisis team was rapidly assembled. Although details remained sketchy, they planned to undertake a major evacuation effort and work began immediately on a whole of government response, including a contribution from the ADF – Operation Bali Assist. As Su watched the news with her husband, John Paul, she theorised how she would organise an evacuation if she had the opportunity and headed off to a conference she was convening that morning at the Royal Darwin Hospital, unaware that she was already on the list of key personnel to be mobilised. The operation was to become the largest Australian aeromedical evacuation since the Vietnam War.[35]

Su's mobile rang at 0930. A Lieutenant Colonel in the Army Reserve, Su was a veteran anaesthetist and intensive-care specialist with extensive military experience. She had served in Rwanda, Bougainville and the Solomon Islands and was a qualified Army Logistics Officer as well as a doctor and anaesthetist. She came from a military background. Her father had served in Vietnam and she grew up longing to go to Duntroon. Instead, she had studied medicine at university and joined the Army Reserve, working as an Intelligence Officer at first. As well as her medical credentials, Su had a strategic mind.

By 1330, Su was aboard an RAAF C130 Hercules with an ADF emergency response team. "I paid off the conference, rounded up all my friends, stole as much equipment and supplies as we could and joined the

---

[35] Hampson, Gregory, Cook, Steven and Frederiksen, Steven, 'Operation Bali Assist: The Australian Defence Force response to the Bali bombing, ', *Medical Journal of Australia* 2002; 177: 620–62.

A new kind of warfare: the rise of global terrorism

first flight."[36] Together with a surgical colleague, Major David Read, she had raided the hospital and a local military facility for supplies, including morphine, antibiotics and surgical equipment. Since accurate information was limited and the team did not know whether diplomatic clearances would allow them to enter Indonesia in uniform, they packed civilian clothes and passports just in case.

During the three-and-a-half hour flight the team wondered what to expect. They knew that a number of Australians were involved, but initial estimations of casualties were low – around 35 – and the aircraft had left Sydney with only four units of blood,[37] sufficient for two patients at best. In fact, Su's team faced a far more serious mass casualty situation. "I was terrified there'd be someone I'd know, as opposed to when I'd been in Africa, or anywhere else," Su recalled. "But you just get there and do the job the best way you can."[38]

Once on the ground in Denpasar, the extent of the bombings became clear. Although the team had expected to pick up the most severely injured casualties from the airport and take them home, there were none there. Instead, the airport was a mass of desperate foreigners trying to flee the country. After acquiring a vehicle and fighting against the tide of traffic, the team headed out to Sanglah Hospital. As Major David Read recalled:

> I still shake my head when I think about it. As soon as we walked in [in Australian Army uniforms] it went stony silent. No one was crying or yelling. They were looking up, with their eyes saying, "Take me home".[39]

Su and her team worked hard to triage patients, continue resuscitation and evacuate casualties by whatever means available. They had to 'reverse triage' where the limited resources available were given to those most likely to survive their injuries. Working conditions were chaotic. While

---

[36] Extract from interview with Su Winter by Sharon Mascall-Dare recorded on 2 December 2013.

[37] Cook, S.

[38] Extract from interview with Su Winter by Sharon Mascall-Dare recorded on 2 December 2013.

[39] Private correspondence between David Read and Susan Neuhaus dated 4 December 2011.

some nurses had written on patients' bedsheets, listing the drugs they had received, other staff had taken the sheets away, removing crucial information that the Australian medical team required.[40]

Soon, media crews began to arrive and crowds of onlookers had to be kept away from the deteriorating patients. Local police and military were called in to help move patients to the airport, where flights were arriving from Australia and neighbouring countries to retrieve patients. The scene became surreal as anxious relatives and friends fought for information and medical staff struggled to cope.

On the tarmac at Denpasar airport conditions were also chaotic. To receive casualties from Sanglah and other hospitals and prepare them for flight, a staging facility was established using the Australians' limited equipment and personnel. With no formal surgical or ICU capability and no pathology facility, the team struggled to manage the high numbers of casualties with severe burns and blast injuries; stocks of oxygen, fluids, antibiotics and resuscitation drugs were quickly depleted.[41]

Indonesia had agreed that the Australians could collect all expatriate casualties to reduce pressure on local resources, and Su found herself dealing with other nationalities as well as Australians. Some of those evacuated to Australia were from Japan, New Zealand, Germany, Scandinavia and the United States. Su played a role in allocating casualties to flights.[42] One incident made her smile amid the chaos:

> A flight came in from Perth … it could take four walking
> wounded. And these four guys who'd been dying on their
> stretchers said, "We can walk!" And they got up and walked on to
> the plane[43].

After several hours' wait, managing 40 patients who were badly burned,

---

[40] Extract from interview with Su Winter by Sharon Mascall-Dare recorded on 2 December 2013.

[41] Cook, S.

[42] After the bombings, casualty movements were released by Sanglah Hospital and posted on the internet. See: http://www.indo.com/bali121002_bekup/mourning1.html

[43] Extract from interview with Su Winter by Sharon Mascall-Dare recorded on 2 December 2013.

A new kind of warfare: the rise of global terrorism

Su returned to Darwin on the second flight back to Australia, supported by newly arrived medical reinforcements. Eventually, 66 Australians were evacuated from Bali, with 22 on Su's flight, all on stretchers. Six of those were seriously injured and two were intensive-care patients requiring ventilation. As the anaesthetist and intensive-care specialist, her priorities were to ensure that each patient received enough fluids and pain relief; she also supervised their treatment throughout the flight. The ventilated patients required constant monitoring and were ventilated by hand throughout the journey.

All of Su's patients were in a critical condition. Burns patients rapidly lose the ability to regulate their own temperature and can lose large amounts of fluid from their burnt skin. Fluids need to be given by drip and regulated closely to avoid causing swelling or organ failure. Limbs can swell rapidly and the burnt tissue acts like a tourniquet, cutting off circulation. With limited space on board a C-130 Hercules, serious burns and blast injuries can be hard to manage. The medical team also had to contend with no heating and constant noise and vibration. "I hate the smell of burns … and these patients were very badly burned. But I was too busy to notice,"[44] recalled Su. "Pretty amazing as the least burned person on my flight was 40 percent."[45]

The aircraft landed in Darwin at 0730 on the morning of 13 October, around the same time that Su had watched CNN the previous day. She headed home to her family, went to bed and returned to work at the hospital that same day.

As a mother, Su had arranged care for her daughter during her 24 hours away. Madeleine was under two. It was the first time that Su had been away since she was born the year before. A friend offered to babysit and Su's husband, a career Army officer, was given permission to miss an exercise and stay in Darwin. Not all his superiors knew the reason why, however:

> About three months later a half Colonel said to him, "Look, you've done a really good job, but we are concerned that you came

---

[44] Extract from interview with Su Winter by Sharon Mascall-Dare recorded on 2 December 2013.
[45] Private correspondence between Su Winter and Susan Neuhaus dated 4 December 2011.

here [to Darwin] for your career, not your wife's and you missed one particular exercise because she had to do something." John Paul didn't have the heart to tell him that "she" went to Bali in uniform with the Army and got a CSC for it.

> I married an Army officer and I've been a good Army wife … but it's always been a balance to try and fit in two careers.[46]

Su was awarded the Conspicuous Service Cross (CSC) in the Bali Tragedy Honours List: "For outstanding achievement in the provision of exceptional medical care to critically injured victims of the bombings as the Specialist Medical Officer to Operation BALI ASSIST, October 2002." Her colleague, Major Read, was similarly honoured, "for carrying out lifesaving procedures at the airport in Denpasar without the normal range of equipment or anaesthetic".[47] The lack of equipment and facilities provided during the operation remained a concern: improvements were made to the ADF's aeromedical evacuation system as a result. The lessons learned were tested once again when there were further terrorist attacks in Bali three years later, in October 2005.[48]

The week after her return from Bali, Su was deployed to East Timor. As fate would have it, the surgeon who accompanied her was none other than Major David Read. Since 2005, she has also completed a full tour of duty in Afghanistan (2008) and was promoted to the rank of Colonel in 2009. While the Army has always been part of her life – her father served in Vietnam – she anticipates a different life for her daughter:

> With an Army dad, an Army husband and being a doctor in the Army, my daughter will probably join the Air Force, or become a pacifist.[49]

---

[46] Extract from interview with Su Winter by Sharon Mascall-Dare recorded on 2 December 2013.
[47] See: http://www.itsanhonour.gov.au/honours/our_honours/ Accessed 16 June 2014.
[48] Cook, S.
[49] Private correspondence between Su Winter and Susan Neuhaus dated 4 December 2011.

A new kind of warfare: the rise of global terrorism

Iraq

Private Vashti Henderson

While the Bali bombings brought the threat of terrorism closer to home, ADF personnel became involved in other operations following the 9/11 attacks, under the broad umbrella of 'the war on terrorism'. From late 2001, Operation Slipper took Australian forces to Afghanistan and the Australian Government deployed some 2,000 personnel to the Middle East in January 2003 to support the US's invasion of Iraq. While the invasion itself began on 20 March and was over within three weeks, Australia was now drawn into a protracted 'nation-building' and counter-insurgency campaign in the Middle East that would see ADF personnel deployed in Iraq until 31 July 2009.[50]

Private Vashti Henderson arrived in Iraq in April 2005, one of 20,000 ADF personnel to serve in Iraq as part of 'Operation Catalyst' between 2003 and 2009. She was a medic with 3 Health Support Battalion (3HSB) – the same unit that Major Suzie Le P. Langlois joined in 1999. Like Suzie, Vashti also served in East Timor – she arrived two years after Suzie's departure, in 2003. As a junior medic deployed outside the capital, Dili, Vashti saw few casualties – she spent much of her time "outside the wire", visiting communities in the countryside. A highlight was working with other ADF personnel and local Timorese to build a church: it was not a hostile environment.

Iraq was at the other extreme. From the moment she entered Iraqi airspace, Vashti knew she was in a conflict zone. Her seat on the C-130 Hercules was made to be functional, not comfortable, and all she could see out of the window was endless desert sand. The aircraft could not land straightaway that day – there was an 'alarm red' and the runway was attracting indirect fire. "We were getting mortared," Vashti recalled, "so we had to do some circles … but it didn't really bother me at all".[51]

---

[50] Australian War Memorial, 'The Second Gulf War 2003–2009' Available at: https://www.awm.gov.au/atwar/gulf/ Accessed 15 January 2014.
[51] Extract from interview with Vashti Henderson by Sharon Mascall-Dare recorded on 7 November 2013.

Lance Corporal Vashti Henderson with General Peter Cosgrove, then Chief of Defence Force, Iraq, 2005. The photograph was taken shortly after General Cosgrove announced her promotion. *Image courtesy of Vashti Henderson*

Vashti continued to demonstrate resilience in the days to come. She had joined the Army Reserve at the age of 17 and was dedicated to her job. Disappointed by the initial training she had received as a medic from the Army, she qualified as an Enrolled Nurse on her return from East Timor and had become a Registered Nurse by the time she went to Iraq. But she didn't want to deploy as a nurse: she wanted to be the best medic she could. She also wanted to test her skills in a war-zone.

It was dark when she arrived on base. Her roommate, an RAAF medic, was working the nightshift and had left a note on Vashti's bunk introducing herself and saying 'G'day'. The next morning Vashti woke up and went to work. From the start, she was surrounded by trauma:

> There were patients the first day. They'd been blown up, limbs missing...
>
> I remember my first patient, his name was Michael Green, and I thought to myself, "Yes, I'm going to remember every patient's name."

A new kind of warfare: the rise of global terrorism

I don't remember a single one after him. There were that many.[52]

On a typical day, Vashti would deal with eight casualties; sometimes she would have seven at one time. Assigned to the Emergency Department of a large US-run military hospital,[53] Vashti dealt with gunshot wounds, IED and MVA (motor vehicle accident) casualties on a daily basis. There were amputations, there were head, chest and limb injuries; patients were usually brought in by chopper and the majority had serious traumatic injuries. Although the medical teams were sometimes "swamped", Vashti found trauma work relatively straightforward: "There's not a lot of hidden stuff – you can guess what is going on with the patient because you know the mechanisms. With medical stuff there can be multiple things wrong with somebody and you have to work it out."[54]

The hospital was relatively well-equipped to cope with the numbers of casualties with four resuscitation bays and six overflows areas; it had three operating theatres that could treat up to six casualties simultaneously. There was an intensive-care unit, where staff were "really overworked", supported by radiology suites and pathology laboratories.[55] Around 25 Australians worked at the hospital, and since the hospital was US-run, they had to adapt their skills accordingly. For Vashti, the sheer number of trauma cases meant she learnt quickly:

> [The Americans] work differently than we do but it's not worse and it's not better. It's just different. We had teams – there was a doctor, a nurse, two medics or an airman and a medic … did it in an American way, where the nurse is actually the scribe.
>
> You'd strip the patient, assess them and shove needles in … and treat what you can see. At times we didn't have enough bays

---

[52] ibid. Green's injuries received coverage in US media. See: http://jacksonville.com/tu-online/stories/042205/nes_18540501.shtml Accessed 16 June 2014.

[53] At the time of writing, the name and location of the hospital remain classified; identifying details have therefore been omitted.

[54] Extract from interview with Vashti Henderson by Sharon Mascall-Dare recorded on 7 November 2013.

[55] ibid.

and towards the end I was team-leading on cases.

I learned so much over there that you just would not have the opportunity to learn in a major trauma hospital here. You just don't see that kind of trauma.[56]

Some staff found it hard dealing with such extreme injuries day after day. While the Americans completed three-month rotations, the Australians were in Iraq for six months, and at times the deaths took their toll. The hospital chapel became a haven for some, offering a few moments respite, and the padre worked overtime offering counselling and support. While Vashti found she was able to cope with the relentless workload and endless stream of horrific injuries, others struggled:

I remember when a contractor came in and he'd been shot in the head – he was dead. We went to roll him and my hand accidentally fell inside his skull.

It was around the same time that the US was swapping over [rotations]. There was this redheaded girl and I didn't think she could go any paler but she did. She got quite dizzy and she had to sit down, poor thing …

It affects people differently. She just needed to get used to it. She got used to it in the end.[57]

Some incidents, however, did take their toll on Vashti. One IED (improvised explosive device) strike hit people she knew – a group of US Army combat medics who used to volunteer at the hospital on their days off. Without warning, Vashti found herself surrounded by friends with life-threatening blast injuries: "I probably had a tougher personality [than others] … but some things over there got to you. When their contingent got blown up we were working on people we knew".[58]

In another incident, she was confronted by the dead body of young soldier who had been shot through the legs. He was brought in by a flight

---

[56] ibid.
[57] ibid.
[58] ibid.

A new kind of warfare: the rise of global terrorism

medic – a young man in his 30s. The medic was tall and strong, "a blokey guy", as Vashti remembers, yet he was close to tears.

> As we worked he trembled with a mix of anger, frustration and adrenalin, I gave him a quizzical look and mouthed the words "You OK?" Nodding, head stooped low, he muttered, "It just gets me when they're so young". I wish I could have said something to help. I tried to think of something to say, but nothing came to me, so I said nothing.[59]

Vashti remembers washing the young man's face; she was struck by how boyish he looked as she prepared his body for repatriation. She stepped outside the curtain, to find the US Army padre standing there in silence, visibly upset:

> He came up to me and said [that] in America it was Mother's Day. This man's mother was probably waiting for flowers from her son. Instead she was going to get a knock on the door to tell her that her son was dead.[60]

Vashti remembers that the padre forced a smile, and said "Thank you" before turning away. She searched for the right words in response, but none came. She resolved to find those words in future, and find a way to console people like other staff members she had come to admire. She also remained determined to become an even better medic.[61]

She continued to work hard for the rest of her deployment and was recognised for her efforts. In June 2005, the Chief of the Defence Force, General Peter Cosgrove, paid a visit to the hospital and asked Vashti to come forward, in front of her colleagues. He promoted her to the rank of Lance Corporal on the spot. "I tried to tell him that he couldn't actually do that, because I hadn't done any courses," said Vashti. "But he was quite

---

[59] Extract from private correspondence between Vashti Henderson and her family dated 8 May 2005.
[60] Extract from interview with Vashti Henderson by Sharon Mascall-Dare recorded on 7 November 2013.
[61] Sourced from private correspondence between Vashti Henderson and her family dated 8 May 2005.

positive that it was in his power and it happened anyway."[62]

On her return to Adelaide – her home town – Vashti continued to build her skills and qualified as a paramedic, working for the South Australian Ambulance Service. She found the transition from full-time Army to part-time reservist and civilian work difficult at first. She felt that others were self-involved and ignorant of Australia's involvement in conflicts overseas. She also felt she had a new understanding of the word 'trauma'.

> I didn't really see anything other than coughs, colds and sore holes before Iraq – we did see a bit in Timor, but it was not really that exciting.
>
> I started as a paramedic straight after I got back … and I remember going to a trauma which, retrospectively, you would consider a significant trauma for South Australia.
>
> At the hospital, they were all talking about this massive trauma and I was just standing there, [thinking] I just don't get it. That really wasn't much of anything.
>
> After seeing a lot of trauma I guess my tolerance levels have changed.[63]

Vashti also feels she learnt a lot about the ethics of conflict medicine in Iraq. While coalition troops had the prospect of follow-up treatment in advanced medical facilities, local people did not have the same opportunity. If they lost a limb or required long-term care, there could be little chance of receiving it locally, in a war-zone. "What's their quality of life going to be if we do everything we can for them, but they get ulcerations from not being moved properly and they die a horrible death?"[64] Vashti was not the first, or the last, to ask such a question. Her response, like others', remains pragmatic: "At the end of the day, we treat everyone the same because everyone deserves the best care that they can get".

In early 2014, Vashti was promoted to the rank of Sergeant and continues

---

[62] Extract from interview with Vashti Henderson by Sharon Mascall-Dare recorded on 7 November 2013.
[63] ibid.
[64] ibid.

to balance her civilian life as a paramedic with her reservist work at 3HSB. Now an instructor to younger medics keen to follow her lead, she remains committed to her Army career as a noncommissioned officer (NCO):

> I've done my nursing degree, but no, I could never be an officer. I like being an NCO: I don't even know what it's going to be like being a sergeant behind a desk.
>
> I prefer the mud.[65]

Lieutenant Hannah Brown

In 2007, Lieutenant Hannah Brown arrived in Tallil, southern Iraq, with a not insignificant task. As the first physiotherapist to deploy to the Middle East with Australian forces, she was tasked with setting up and running the first physiotherapy department in the Middle East Area of Operations (MEAO).

At the time, Hannah had less than one year's experience as a practising physiotherapist and army officer. She had joined the Australian Army Reserve as a student in 2002, initially in the Royal Australian Corps of Transport. In her third year of study, the Army granted her an undergraduate scholarship to complete her studies in physiotherapy – in return, she would join the ADF for a minimum of three years' full-time service on graduation.

On her arrival in Iraq, Hannah realised that she had a lot to learn in a short time. She describes those early weeks as a 'process of trial and error':

> I decided to take on postgraduate studies while deployed in an attempt to fast track my clinical knowledge. The study and a supportive medical team help significantly, but the learning curve was overwhelming. I learnt a lot about my trade during this deployment, particularly the physical, mental and emotional strains that deployed soldiers face. I saw injury patterns very unique to the army and began to realise their complexity.[66]

---

[65] ibid.
[66] Brown, Hannah, private email correspondence to Susan Neuhaus, 30 January 2010.

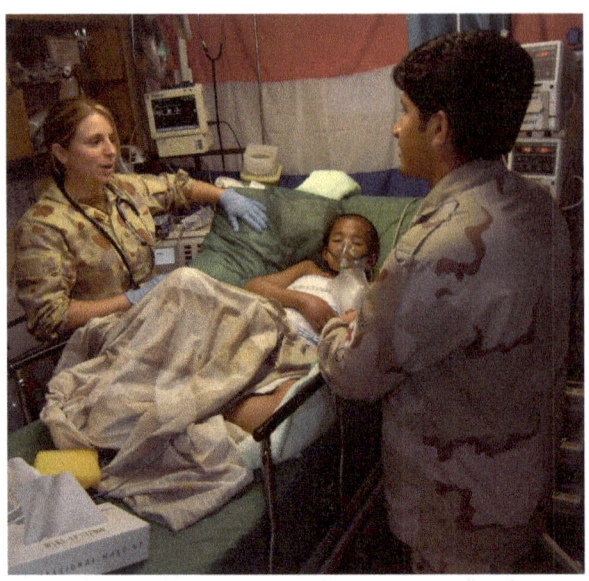

Lieutenant Hannah Brown providing physiotherapy services to a child in the NATO Role 2 health facility, Afghanistan 2009.

Twelve months after her return from Iraq, Hannah deployed to Afghanistan and spent seven months based in Tarin Kowt. This time she felt more prepared for the role. Concerned that physiotherapists had not been adequately recognised as having a core capability in the 1990s, she did her part to contribute to greater recognition for her profession:

> When large Australian Army contingents began to deploy overseas in the late 90s the concept of deploying a physiotherapist was viewed by sceptics as 'a waste of a position' or an 'unnecessary luxury'.[67] Since this time it has been a battle for military physiotherapists Australia wide to justify their valuable contribution as a force multiplier to a deployed battle group. The work done overseas by physiotherapists, past and present, now speaks for itself. Physiotherapists have been successful in reducing

---

[67] These observations reflect the experiences of previous military physiotherapists. Sir Neville Howse VC, Surgeon-General in 1914, described inclusion of masseuse services 'a waste of time'. (see chapter 2). By World War II their role became more established (see chapter 4), but in Vietnam, Captain Skewes reports similar observations (see chapter 5).

the number of soldiers experiencing chronic injury, reducing recovery time and even preventing soldiers from being medically evacuated for the Area of Operations.[68]

In 2009, Hannah was promoted to the rank of captain and now oversees new physiotherapy graduates entering the ADF.

Captain Alison Malpass

Some 20,000 members of the ADF served in Iraq as part of Operation Catalyst from 2003–2009,[69] supported by dedicated teams of Iraqi interpreters and other staff. As the operation drew to a close and Australia began to bring its people home, the safety and security of local staff became a priority.

Revenge attacks on Iraqi interpreters were endemic and there was pressure on Australia – alongside the United States, the United Kingdom and their partners in the 'coalition of the willing' – to grant special protection visas to Iraqis who had supported coalition forces. The List Project to Resettle Iraqi Allies – an advocacy group backed by a number of high-profile US law firms – estimates that more than a thousand Iraqi interpreters were murdered in revenge attacks, with many more abducted, tortured or injured during combat.[70]

In order to grant refugee visas and government transport to Iraqi interpreters and their families, Australia adopted a 'Whole of Government' approach, with an ADF medical screening team working alongside teams from the Departments of Defence and Immigration.[71] Called the ADF Supplementary Team, the contingent also included command and control, policy, administrative, legal, linguistic, logistic and communications personnel. To protect the identity and security of those involved, the operation was highly classified.

---

[68] Brown, Hannah, private email correspondence to Susan Neuhaus, 30 January 2010.

[69] See: http://www.defence.gov.au/opex/global/opcatalyst/index.htm Accessed 16 June 2014.

[70] See: http://thelistproject.org/new-website-new-list/Accessed 16 June 2014.

[71] Duffy, Peter, 'On the Ground in Iraq', *Inside News*, Royal Australian and New Zealand College of Radiologists, Vol. 5 No. 1, December 2008, pp. 24–5.

Corporal Shannara Curran, Sergeant Michael Kopacz and team doctor Captain Alison Malpass (right), liaise with a locally engaged employee during a health assessment interview.
Image: Department of Defence

The team deployed from April to June 2008 and was led by Major Caitlin Langford, RAAMC. Leading and coordinating its 29 members of staff from diverse professional backgrounds was a feat in itself – there was no template or 'script' to work from in meeting the requirements of the mission. When the initial tasks – checking applicants' eligibility for visas and health screening – were complete, Major Langford had another challenge. Unexpectedly, she was given the task of leading the extraction phase, where entire families were subsequently transported safely from Iraq to start a new life in Australia.

> It was a challenging mission … it offered an opportunity for me to lead a group of people who had not worked together before. It required detailed planning and consideration of all aspects. How, at a tactical level, do you get families out of the country safely who are potential targets? Some of the family members were infants; some were elderly. They were all clearly anxious, yet hopeful, about what lay ahead.

A new kind of warfare: the rise of global terrorism

> The extraction phase saw every team member extend themselves and work outside of their core role ... The team always treated the Iraqis with dignity, considering all aspects of their health, welfare and culture. The strength and resilience of the Iraqis was amazing during what must have been an overwhelming prospect – packing your life into two bags and heading to the southern hemisphere for a new life.
>
> Our mission was underpinned by a strong moral sense this was the right thing to do for these Iraqi families who, by virtue of their support to the ADF and its mission, had risked their and their family's own personal safety. It was a privilege for all of us to be part of the process that offered them a better life in Australia, and for me it was a career highlight to lead such a positive team.[72]

Alison Malpass was a doctor and reservist; she was sent to southern Iraq as part of the ADF Supplementary Team. After joining the Army as a reservist in 1997, she had qualified as a doctor in 2003 and was commissioned as a Specialist Service Officer in 2007. When the opportunity arose to go Iraq she didn't hesitate: she saw it as an ideal opportunity to test her skills on a 12-week deployment overseas:

> When my CO [Commanding Officer] called me about the deployment ... there was never a decision to be made. It was my obligation to go. I'd been a reservist all these years, now they were asking me to go and do a job, so it wasn't really an option. That was how I felt ... a sense of duty.[73]

The team lived in tents and worked in converted shipping containers – both were airconditioned but struggled to offer respite from daily temperatures that regularly hit 45°C. Over 300 medical examinations were completed; radiologists also took 165 chest X-rays to screen visa applicants for tuberculosis. This was a requirement of the Australian immigration

---

[72] Extract from interview with Lieutenant Colonel Caitlin Langford by Sharon Mascall-Dare conducted on 27 June 2014.
[73] Extract from email correspondence between Sharon Mascall-Dare and Alison Malpass dated 21 November 2013.

process, along with pregnancy testing and confirmation of 'fitness to fly'.[74]

Some of the work was repetitive to some extent and there were few surprises – fortunately, no cases of tuberculosis (TB) were found, as Iraq had similar rates of TB infection to Australia. As the only female doctor in the team, Alison's key role was to screen the wives and daughters of Iraqis who often knew nothing about their husband's work. For some women, it took years before they knew the full extent of their husband's involvement – security of information was essential throughout the operation. Without that security, lives were at risk.

High security also prevented staff from leaving the base – no-one was allowed to go beyond the perimeter. As a result, Alison saw very little of the surrounding countryside and her contact with Iraqis was confined to her role as a doctor. At times she felt she was "wrapped up in cotton wool" as her freedom was restricted to ensure her safety. As a female, her contribution to the success of the mission was crucial:

> It was very important for the male interpreters to know that their wives were going to be treated in a way that was appropriate for their religion and their culture. It was important that their wives were examined by a female.
>
> Where we worked was between the inner and outer perimeter of the base – you could go and peer out into 'no man's land'. But I wasn't allowed to go. "What if Alison got shot", the others would say, "We can't lose Alison." I wasn't even allowed to go to the gym on my own.
>
> The CSM [company sergeant major] wanted to make sure I was safe … but it wasn't anything to do with my being female. It was just that they needed me in the team. Without a female doctor they couldn't do the examinations.[75]

Although she was kept safe, there were moments when Alison was reminded that she was in a warzone. A missile landed just 100 metres from

---

[74] Duffy, Peter, 'On the Ground in Iraq', *Inside News*, Royal Australian and New Zealand College of Radiologists, Vol. 5 No. 1, December 2008, pp. 24–5.

[75] Extract from interview with Alison Malpass by Sharon Mascall-Dare on 2 November 2013.

her on one occasion, causing an explosion that fired sparks up into the sky. "It was a like a fireworks display at first," said Alison. "Then we knew what it was. I didn't feel in danger, but it was a reminder".[76]

After Iraq, Alison went on to work as a locum doctor in the United Kingdom, specialising in emergency medicine and continuing to work in a multicultural context. "In Iraq, I learnt that a smile can get you a long way", she recalls. "Being able to win women's trust so that they were prepared to disrobe to their underwear, and I could listen to their chest; I learnt a lot from that. Interacting with other cultures is something I really enjoy."

Defence formally concluded its military commitment to the rehabilitation of Iraq on 31 July 2009, with the final 11 ADF members, working in various United States-led coalition headquarters, withdrawing on 29 July 2009.[77]

Captain Alison Malpass was later recognised for her contribution to the mission with a Chief of Defence Force Group Commendation. Her response was self-effacing. "I just did my job," she said.[78]

Afghanistan

Corporal Jacqui de Gelder

As Operation Catalyst came to a close, the war against the Taliban in Afghanistan was still ongoing – Australia's commitment to Operation Slipper in the Middle East Area of Operations (MEAO) would continue for at least another five years.[79] The war in Afghanistan has become the longest involving Australian Defence Force personnel – exceeding the durations of World War I and World War II combined. It defined a new kind of warfare at the start of a new millennium – the enemy was armed with simple but devastating technology as well as ideology. These combined to form the

---

[76] ibid.

[77] See: http://www.defence.gov.au/opex/global/opcatalyst/index.htm Accessed 16 June 2014.

[78] ibid.

[79] At the time of writing, Australian troops are set to withdraw from Afghanistan by the end of 2014.

signature injury of the Middle East conflict – the improvised explosive device (IED).

Corporal Jacqui de Gelder dealt with her first IED blast on her first patrol. She had just joined the Australian Mentoring and Reconstruction Task Force in central Afghanistan, where she was deployed as a medic. The work was dangerous and unpredictable – during her eight-month tour of duty she dealt with traumatic amputations, blast injuries, burns and the aftermath of a suicide bombing. Her skills were tested from that first IED, on her first day of patrolling.

Fortunately, Jacqui had experience to draw on. She had served in Kuwait and Iraq for four months over the Christmas period of 2007–8, conducting medical checks on personnel deployed under Operation Catalyst. Shortly before her deployment to Afghanistan she had also been exposed to a new kind of medical 'training'.

On 9 February 2009, her older brother, Paul, was attacked by a shark in Sydney Harbour. He was a navy clearance diver: Jacqui had three brothers and they all served in the ADF. Her family tried to contact her immediately, but she was away from her phone attending PT on her base in Townsville. Once the message got through, the Army put her on the next flight to Sydney.

> It took my breath away. There were already so many people in the ICU … I went in and Paul had tubes down his throat, he was a bit in and out of consciousness. At the time he still had his leg but his hand was missing. It was horrible.[80]

Paul lost his leg as well as his hand following the attack. Although Jacqui was only two weeks into her posting in Townville, she was given compassionate leave to stay in Sydney and was eventually offered a detachment to the Submarine Underwater Medical Unit at HMAS *Penguin*. She worked closely with navy clearance divers, like her brother, and drew inspiration from his survival. With hindsight, Jacqui believes her brother's accident helped to prepare her for Afghanistan.

---

[80] Extract from interview with Jacqui de Gelder by Sharon Mascall-Dare recorded on 17 December 2013.

A new kind of warfare: the rise of global terrorism

Corporal Jacqui De Gelder relaxes with an Afghan colleague while working with Mentoring and Reconstruction Task Force 2 (MRTF2). *Image: Department of Defence*

Jacqui was deployed at short notice to Afghanistan, and had to move fast to get her vaccinations up-to-date. She was sent to Kuwait for a week of training that involved briefings, weapons drills with body armor and how to identify an IED. As a medic, she was also trained to use emergency equipment required in the field, including self-applied tourniquets, chest tubes, tracheotomies and intraosseous (IO) drills, which injected fluids direct into bone marrow. This training proved to be vital.

Her first patrol began without incident. She was based in Tarin Kowt, the capital of Oruzgan province, and was patrolling in the Baluchi Valley, between Tarin Kowt and the town of Chora. It was an area that the Taliban wanted to control. Firefights between the Taliban and coalition forces were common, and there have been significant Australian casualties.[81] Surrounded by mountains and desert, the valley led into a 'green zone' of cultivated farmland, aquaducts and a river. While some of the local people close to the patrol bases were friendly, it remained dangerous territory.

---

[81] The Department of Defence lists Australian deaths in Afghanistan on its website: http://www.defence.gov.au/vale/ Accessed 6 February 2014. The ABC reports the location of each death, including those in Oruzghan Province and the Chora Valley: http://www.abc.net.au/news/2011-05-25/australian-casualties-in-afghanistan/1784808 Accessed 6 February 2014.

Jacqui was around 200 metres away from the IED when it exploded. She was sitting inside a traditional dwelling – known as a quala – waiting for the area to be cleared. She had not been assigned medical duties that day – as one of the few females working with the infantry, she was carrying out searches on local women. It was 'hearts and minds' work, where relations were built and maintained with the local community. The women would be searched in private, to respect local custom, and Jacqui would provide them with medical assistance if they needed it.

Within seconds of the blast, the news came through on the radio. "We've got medics here. We can send them up," said her patrol commander, and Jacqui headed towards the site of the explosion. Private Benjamin (Benny) Renaudo, of the 1st Royal Australian Regiment (1RAR), had been killed instantly[82]. His friend, Private Paul Warren, had lost his leg. It was Jacqui's first major trauma:

> When we got there, the engineers were still trying to clear a path to get to the casualties because they weren't sure if there were any more IEDs around.
>
> One of the other medics … said that Benny was dead; he died instantly … our main priority was to get Paul Warren out of there and back to Tarin Kowt because he had a traumatic amputation to his leg.
>
> Before I treated that first big trauma, I'd never had a 'proper' priority one injury during my time in Australia. I was a bit worried about it because I didn't know how I'd react to seeing a big injury, or how I'd go as a medic, doing my job in that situation. But, after that first one, I felt pretty confident within myself and in my own skills.[83]

Once a path to the casualties has been cleared, Jacqui put a second tourniquet on Paul Warren's leg – the first was applied by a combat first-aider, who was already on site. They gave Paul a 'green whistle', with the

---

[82] See http://www.defence.gov.au/vale/pte_ranaudo/pte_ranaudo.htm Accessed 7 February 2014.

[83] Extract from interview with Jacqui de Gelder by Sharon Mascall-Dare recorded on 17 December 2013.

A new kind of warfare: the rise of global terrorism

pain-reliever methoxyflurane, and struggled to insert an intravenous line, since he had gone into shock. To get fluids in, the team used an intraosseus drill into Paul's shoulder. Once stabilised, he was airlifted out by helicopter.

During the eight months of her deployment Jacqui dealt not only with a suicide bombing in the Chora Valley, but with countless trauma cases at her patrol base in the town of Mirwais or the main hospital in Tarin Kowt. Nothing, however, compared to that first 'big trauma' on her first patrol. Her actions proved she could do the job – to herself and the men around her:

> None of the boys really spoke to me or accepted me at first. But after the first incident … I was fine with the guys. It's like that. You just have to earn their trust.
>
> Once they know that you're good at your job then you're fine, no matter female or male. If you're not a good medic being a male then they're not going to accept you. If you're not a good medic as a female, it's going to be the same. So I don't think sex came into it. It was more, "Can you do the job?", "Do we want you as our medic?", "Can we trust you?"[84]

Jacqui still found some aspects of the job difficult as a female. Wearing body armor, webbing and a helmet while carrying medical equipment and a weapon was no easy task. Although Jacqui was physically fit, her duties in Afghanistan required a different kind of fitness. At times her shoulders would get sore from the weight; sometimes the Afghan interpreters would help carry the 'med kit'. Jacqui also found ways to reduce her load, but keeping airway equipment and haemorrhage control in her webbing.

Jacqui's height was also a drawback. The patrols would cross rivers, and while the men were wet to their knees, the water would reach the top of Jacqui's thighs. In summer it kept her cool; in winter it was uncomfortable. To keep going, she would think about her brother, Paul, who was still recovering from the shark attack back home. "If I was tired from patrolling then I used to just think of Paul and think, "OK. Well look what's he's

---

[84] Extract from interview with Jacqui de Gelder by Sharon Mascall-Dare recorded on 17 December 2013.

done, he's got his leg cut off and then the next day he was up and walking. If he can do that, I can continue on."[85]

A trickier problem was the toilet. With no privacy, and long journeys in armored vehicles, Jacqui found it hard to find somewhere to 'go'. While other soldiers were worried about hitting an IED, Jacqui had other concerns:

> Where can you go when you're in the middle of the desert so no one can see you? I went out in a Bushmaster – it was for 10 hours – and it was my first trip so I was really paranoid about going to the toilet. I didn't drink any water and I don't think I went to the toilet. I had the worst headache at the end of the day and I thought, "No, I can't continue like this. They're just going to have to get over it. I'm going to have to get over it", and you know, by the end of the trip, no-one even cared. It was, "Yeah, Jacqui's gone to the toilet." You just have to deal with it and get over it.[86]

At one point Jacqui spent a few weeks with a platoon who had rented a quala. It was a highlight of her deployment, even though she still had little privacy. She would heat up bottles of water in the sun and have a makeshift shower while the men were out on patrol. The 'toilet' was a plastic bag and a wooden seat: "I had to use my hootchie[87] because I was the only one who wanted privacy, but the boys were respectful and in the end they got used to it, even though I did get busted a couple of times … very embarrassing."[88] On another occasion, she was 'caught short' out on patrol:

> This one time on a mounted move I had to urinate, so I had to ask the guys to stand out of the hatch of the Bushmaster while I did my business in a TravelJohn. A TravelJohn is a small bag that turns your urine into jelly. While I was urinating the boys

---

[85] ibid.

[86] Extract from interview with Jacqui de Gelder by Sharon Mascall-Dare recorded on 17 December 2013.

[87] A hootchie is a thin tarpaulin used in the Army to make a shelter.

[88] Extract from interview between Jacqui de Gelder and Susan Neuhaus, 12 December 2011.

were teasing me about not peeing on the floor, so I contemplated starting a jelly fight – with the contents of the TravelJohn.[89]

While Jacqui felt she was treated like "one of the boys", the Australian media singled her out for being female, calling her 'GI Jacqui'[90] and used her photo to highlight controversy about 'Lady Killers'[91] and combat roles for women on the frontline. Her mother was surprised – she opened the newspaper one day to see Jacqui's photo, complete with webbing and weapon – but Jacqui did not react: "I think they're always saying, 'First Female on the Frontline', but there's been a lot of females that have been in combat. I'm not the only one."[92]

Jacqui's gender attracted interest from local people in Afghanistan, as well as the media back home. Young girls would seek her out, fascinated by the fact she was a female with blonde hair, something they had rarely seen. She earnt the respect of local women, who would turn to her for medical advice. Afghan soldiers also saw her as a novelty:

> I never had a problem being a female with the locals or the Afghan National Army. I thought it was going to be a problem, but if anything, they liked me being a female more.
>
> There was one incident – I was a medic at Patrol Base Wali providing medical support... One of the Afghan soldiers was complaining of a stomach-ache and I felt his stomach. I was doing my assessment and then the next one came up and said the same; and then, the next one came up and he had the same problem. I worked out they all just wanted me to touch their stomach. I said, "OK. That's enough."
>
> [Once] the Australian platoon commander had to ask the

---

[89] ibid

[90] McPhedran, Ian, 'GI Jacqui Gelder packs heat in Afghanistan', the *Telegraph*, 10 September 2009, available at: http://www.dailytelegraph.com.au/gi-jacqui-gelder-packs-heat-in-afghanistan/story-e6freuy9-1225771265565 Accessed 7 February 2014.

[91] Summers, Anne, 'The Lady Killers – Women in the Military', *The Monthly*, December 2011, p. 24–7, available at: http://www.themonthly.com.au/issue/2011/december/1322725684/anne-summers/lady-killers Accessed 7 February 2014.

[92] Extract from interview with Jacqui de Gelder by Sharon Mascall-Dare recorded on 17 December 2013.

> Afghans for permission for me to come on patrol because they didn't want to offend them by saying that a female could do their job. But, as far as I knew, they were happy. When we crossed one of the points in the aquaduct, an Afghan soldier held my hand crossing it.[93]

Jacqui left Afghanistan in February 2010. She returned to Townsville and was later posted to the Royal Military College – Duntroon. On Australia Day 2011, she received a Commendation for Distinguished Service, recognising her "medical knowledge and for skill, courage and selflessness while in Mentoring and Reconstruction Task Force 2, on Operation SLIPPER in Afghanistan in 2009".[94] Having achieved so much, she is now studying at university in Perth, hoping to work for the Royal Flying Doctors Service. For Jacqui, it is now time to move on, and continue her career outside the Army.

> For me, I feel like I achieved everything that I could, and I don't feel like I can achieve much more. I have reached the rank of sergeant … when I joined I never thought that I would get to [that rank].
>
> I've had two deployments; to me I don't feel like there's much more that I can achieve. That's why I want to go to university to be able to achieve more.[95]

OTHER MEDICAL WOMEN IN THE MIDDLE EAST AREA OF OPERATIONS

Women also took on other roles as medical professionals during the 'War on Terror'. This included a surgeon (Colonel Susan Neuhaus) and two female anaesthetists: Lieutenant Colonel Su Winter and Lieutenant Colonel Fran Smith.

---

[93] ibid.
[94] Citation available at: http://www.itsanhonour.gov.au/ Accessed 7 February 2014.
[95] Extract from interview with Jacqui de Gelder by Sharon Mascall-Dare recorded on 17 December 2013.

A new kind of warfare: the rise of global terrorism

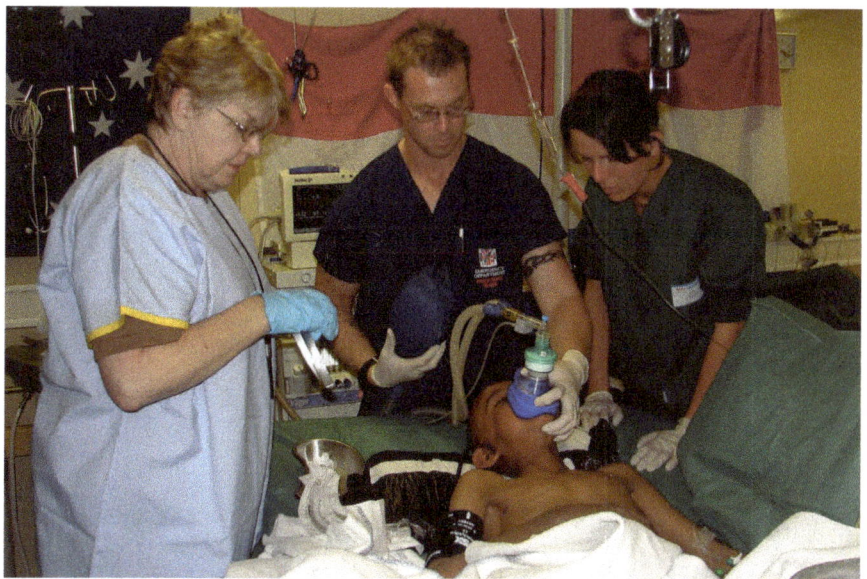

Lieutenant Colonel Fran Smith, specialist consultant anaesthetist and anaesthetist working with the intensive care team Role 2 NATO health facility Uruzgan. *Image courtesy of Colonel Richard Mallet*

Fran Smith deployed as an anaesthetist as part of Australia's medical contingent in Afghanistan in early 2008. She was part of the first of four Australian Medical Treatment Force (AUSMTF) rotations, which augmented the surgical capacity of the Dutch Role 2 Medical Treatment Facility in Tarin Kowt. This was Australia's first and main commitment of surgical and intensive-care personnel into the Afghanistan area of operations. AUSMTF-1 completed 67 operative procedures on 45 casualties – the majority of cases were Afghan local nationals and many required two or more operative procedures.

Fran was uniquely qualified as a physician, anaesthetist and intensivist as part of this team. Combined with her previous operational experience, she was able to make a tangible contribution to post-operative care and improving patient outcomes. She was often required to follow patients from resuscitation, through damage-control survey and to recovery in intensive care. It was intensive, challenging work. It required working closely in a bi-national team, with sometimes quite different views on patient management. Many of the casualties were children and this required

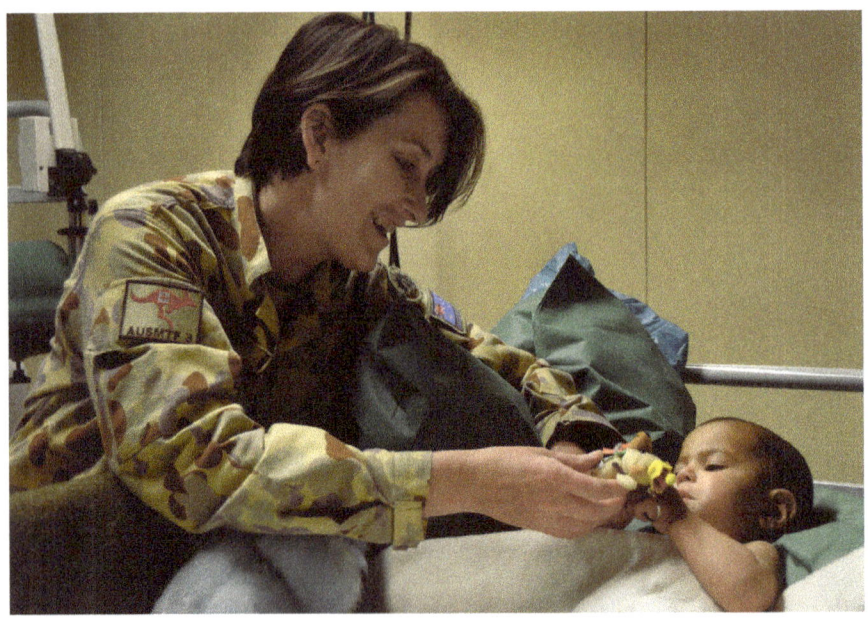

Colonel Susan Neuhaus, Clinical Director Role 2 NATO health facility Uruzgan, Afghanistan with child, 2009.

considerable improvisation, particularly as the ventilators and anaesthetic equipment was designed for adults. The weather was exceedingly hot. The days were long and arduous and always there was the prospect of major battle casualties. Despite such demands, Fran was also responsible for helping to implement a new system, known as, ADF SURG ASSIST, to enable the specialist teams, for the first time, to track their own clinical outcomes and formalise the lessons learned in this new type of warfare.

CHAPTER EIGHT

Now and the future: women and medical command

In too many cases the team has been defined through exclusion of women. This simply has to stop ... Organisations with high levels of what can be termed as 'social capital' are more effective, both in their performance and ability to retain their highly skilled personnel much longer.

In other words making the most effective use of our female soldiers makes good sense. It enhances our capability. That is a simple truth".

Lieutenant General David Morrison AO, Chief of Army, United Nations International Women's Day Conference, 2013[1]

They talk about supporting women to remain in the ADF, they talk about supporting women to reach senior ranks, they talk about of affording women the opportunity to demonstrate the skills necessary to reach senior ranks. I'm one of those women ...

Brigadier Georgeina Whelan AM CSC, 2013.[2]

Warrant Officer Class One Tracey Connors

The smiles in the back of the C-130 were disconcerting. Sergeant Tracey Connors looked around at the men's faces, determined not to give herself

---

[1] Speech. United Nations International Women's Day Conference, New York, 8–9 March 2013. Available at: http://www.army.gov.au/~/media/Files/Speeches/CA_UNIWD_8-9MAR2013.pdf Accessed 16 June 2014.

[2] Whelan, Georgeina, extract from interview with Sharon Mascall-Dare on 2 November 2013.

away. It was a crisp, clear day above the coast of New South Wales, and the Parachute Surgical Team (PST) was preparing to step off. For Tracey, it would be her first jump.

Physical fitness was not an issue. Standing almost six feet tall and "naturally fit", Tracey found the Army's physical requirements relatively easy – she had thrived as a recruit at Kapooka, unfazed by the 'face-ripping', where a corporal would bawl her out inches from her nose, or a sergeant would push her to run further and faster than the day before. When some recruits struggled, Tracey kept going. Army life suited her: she had a love of shooting and the outdoors.

Still, as she sat in the back of the aircraft that day, circling the landing site at 10,000 feet, Tracey was outside her comfort zone:

> I was fit enough – that was never an issue. But I was not a happy camper on my first jump. I was scared like everyone else and it was funny, because you'd sit in the aircraft looking at the guy on the other side and everyone would be smiling.
>
> They were probably crapping themselves more than you, but all you could do was sit there thinking, "Oh my God, oh my God!"[3]

As she landed safely, Tracey had made her mark. She was one of only three women posted to the PST that year and was set to achieve even more in her military career. In 2008, Warrant Officer Class One Tracey Connors would become a Regimental Sergeant Major of the Royal Australian Army Medical Corps.[4]

Tracey is candid about her journey into the ADF. Her childhood was not an easy one. Her father served in the Air Force and the family moved around frequently. Tracey was the second child of four. She was born in

---

[3] Connors, Tracey, extract from interview recorded with Sharon Mascall-Dare on 29 October 2013.

[4] Tracey Connors is one of very few women to reach the rank of RSM in the RAAMC. The first was Warrant Officer Class One Marion Bowen, appointed RSM of the School of Army Health in 1995. WO1 Bowen also became RSM of 2HSB and Corps RSM. See: Morgan, Barry, 'School of Army Health – Home of the Corps: A Brief History of the School of Army Health', p. 7, available at: http://straskye.tripod.com/pages/documents/soahhistory.pdf [accessed 26 June 2014].

Wagga Wagga, while her younger sister was born in Adelaide and her brother in Queensland. When she was 10 years old her father left, and her mother brought up four children on her own.

Money was often tight. Tracey would see her father once a fortnight. She maintained more frequent contact with him after joining the Army. As rebellious as she was resilient, Tracey recognises that she grew up "harder" than her peers. Once she arrived at Kapooka, her upbringing had its advantages: Tracey took the yelling, 'face-ripping' and discipline in her stride.

She enlisted in the Army at the age of 22. After some initial reservations, her mother encouraged her to join, concerned that her daughter had fallen in with the 'wrong crowd'. "She grabbed me by the ear, chucked me down the recruiting office and said, 'Get in!'" Tracey recalls. While her father's career was also a factor in her career decision, "Defence was always in the background,"[5] another major influence was Tracey's younger brother, who was an officer in the Army cadets. He explained the workings of the Army's SLR (Self-Loading Rifle) and taught her basic navigation skills. His guidance gave her a head start.

After her enlistment in February 1991, Tracey began recruit training at the 1st Recruit Training Battalion (1 RTB), Kapooka, in New South Wales. At the time the recruit course was segregated, with females and males allocated to separate platoons. The course involved 16 weeks of arduous physical training, field-craft, navigation, first aid and weapons training in addition to lessons and drill practice. From the outset, Tracey's aptitude was clear – "I didn't have any issues – I could learn, I enjoyed it because I was very outdoorsy. I enjoyed shooting."[6] She also demonstrated signs of leadership – qualities that would come to the fore as she pursued her career as a member of the RAAMC.

After graduation from Kapooka, Tracey completed her Initial Employment Training (IET) as a Medical Assistant at Portsea, Victoria. It was not long, however, before she returned to her 'alma mater' – the Army's school of soldiering, this time as an instructor. Kapooka, with

---

[5] ibid.
[6] ibid.

its ethos of drill and discipline, became her home for the next few years. Now a corporal, she had found the ideal place to pursue her ambition of professional soldiering and rapidly completed her promotion courses. By the time she left, she had been promoted to sergeant:

> Kapooka stands out for me because it gave me a really good knowledge base, which set me up for the rest of my career.
>
> You're counselling soldiers; you're learning how to write reports at quite a young age, as a corporal. I could do my trade as a medic … but I could also pick up mistakes in people's dress, their drill and small-arms coaching.
>
> I picked up a lot of experience and worked in an all-corps environment.[7]

It was experience that served her well in her next posting – as an instructor at the Australian Army's elite academy: the Royal Military College – Duntroon. Here, she was responsible for training and moulding young men and women into officers of the Australian Army. Although she was often the only female on staff – as had been the case at Kapooka – her gender was no barrier:

> I worked for a fellow called Paul Burns – he was my captain. That guy was an absolute legend. He was from Special Forces and he taught me that even though I was the only female, and I didn't necessarily have much knowledge about infantry tactics, that I still had something to offer.
>
> During the round table discussions, he'd ask the infantry guys, the armour guys and the others who had experience in IMTs or Infantry Minor Tactics. And he'd say, "Trace, what do you think?"
>
> I learnt that as a leader, you need to be inclusive and everyone has something to offer.[8]

Tracey's next posting was a case in point. She joined the 1st Health Support Battalion (1HSB) as the Company Headquarters Sergeant at a

---
[7] ibid.
[8] ibid.

Now and the future: women and medical command

Warrant Officer Class One Tracey Connors.

time when the unit was amalgamating with the elite Parachute Surgical Team (PST). For several years her colleagues had flagged her suitability for the PST, acknowledging at the same time that women were not allowed to join. But by the time that Tracey joined 1HSB in 2001 the PST's eligibility requirements had changed. Women's contributions were being acknowledged and their roles were changing, accordingly.

The PST was a highly trained, elite surgical team. In July 1998, three years before Tracey's posting, the PST deployed to Vanimo, off the northern coast of Papua New Guinea after a tsunami devastated the region. The ADF was tasked to spearhead the humanitarian assistance effort, known as Operation SHADDOCK, and within 48 hours, members of the 1st Parachute Surgical Team (PST) were in place.[9]

The surgical workload was extremely high in the first several days, with up to 150 patients arriving at a time. Two hundred and nine procedures were performed in just 10 days. Most of the initial workload was debridement

---

[9] Taylor, P. R. F., Emonson, D. L., Schlommer, J. E., 'Operation Shaddock – the Australian Defence Force response to the tsunami disaster in Papua New Guinea', *Medical Journal of Australia*, 1998; 169: 602–606.

and amputation – some 14 amputations were performed. After day five the workload was dominated by delayed primary closures and grafting. A quarter of the patients were children.[10]

At Vanimo, the PST had demonstrated its ability to deploy at short notice in an area that was difficult to access. Typically, the team comprised up to 30 members, who carried surgical instruments, tools and packs of fluid; the bulkier equipment, including resuscitation, X-ray and pathology gear was delivered via heavy drop. Supplies were 'cross-loaded' between PST members on different aircraft, ensuring a basic team could function if one or more aircraft went down or individuals were unable to reach the site. The team would establish a surgical facility to perform initial wound surgery under canvas; one area was assigned to X-ray and pathology, and another to surgery. Once deployed, the first task for the PST was to get back to the drop zone, clearing casualties on the way. Depending on case load, the unit could manage for three days before resupply was required.[11]

The PST came under the banner of the 3rd Royal Australian Regiment or 3RAR. At the time, the unit had a reputation as a 'boys club' – it was male dominated, in comparison to other units. At the turn of the millennium, however, the culture was changing and female interlopers were finding acceptance. This did not mean changes in behaviour: there was still swearing and no allowances were made for the women who deployed alongside them. Respect had to be earned. Tracey and her female colleagues had to prove themselves first.

As the Company Sergeant Major of the Parachute Surgical Team (PST), Tracey made 12 or 13 jumps during her posting. Credibility came through not flinching from the hard physical standards and route marching with the men for long distances carrying the loads that were expected from everyone.

In 2004, 3RAR's Commanding Officer issued wings to all members of

---

[10] The PST deployed two operating tables, one general/vascular surgeon and one orthopaedic registrar. The team was later reinforced by two orthopaedic surgical teams, one from Monash Hospital and the other from the RAAFSR. See: Neuhaus S, 'Post Vietnam – three decades of Australian military surgery', *ADF Health*, 2004; April 2004;5(1):16–21.

[11] Connors, Tracey, extract from interview recorded with Sharon Mascall-Dare on 29 October 2013.

the PST, including Tracey. She wore her wings with pride:

> It raised a few eyebrows when I wore my uniform to the mess or when I was posted as a career manager. The infantry guys would say "Hey, what's the girl wearing our wings for?"
> The RSM would say: "Shut your mouth. You don't know what you're talking about. She's done more jumps than you, so – on your bike!"[12]

Tracey also earned the Advanced Marksmanship Badge (known as the 'Crossed Rifles') – a benchmark for the Army's most proficient marksmen and women. Confident in her abilities, yet modest about her achievements, she attributes her success on the day of assessment to sufficient rounds of ammunition and time to practise. Worn together, on her uniform, her wings and cross rifles are powerful symbols – as she worked her way up the ranks they have helped to her gain respect, especially from her colleagues in the infantry or arms corps.

Tracey has also gained respect on deployment. From February to August 2002, she deployed to East Timor on Operation TANAGER and Operation CITADEL; in 2004, she deployed to the Solomon Islands as part of Operation ANODE. There, during training drills, her physical fitness was tested:

> I was on an operational patrol with a section from 6 RAR whilst deployed as a WO2 in the Solomon Islands in 2004. I was expected to carry the same weight, march the same distance as well as be a member of a section.
> Although the section commander was a corporal I had no trouble with being integrated … in fact, all of the section members were very inclusive.
> We covered 36 km in three days. I am sure they enjoyed watching me during the conduct of the pre-patrol rehearsals, especially during the contact drill i.e. run, down, crawl, observe, aim, fire – crawling on my belly across the jungle.[13]

---

[12] ibid.

[13] Connors, Tracey, extract from private correspondence with Susan Neuhaus dated

Unlike other members of the RAAMC who were largely confined to ship during their deployment to the Solomon Islands,[14] Tracey was deployed as a medic, working in outposts, on shore, as a part of the 1HSB surgical contingent. After flying into Honiara, she spent a week with the environmental-health team before heading out to Auki, the capital of Malaita province, by helicopter. The breakdown in security was immediately apparent – she was greeted by a local murderer on her arrival – he was sitting at the airfield with his friends, waiting for transport to Honiara and its overcrowded prison. She then travelled on to Atori, where she would swim with children from nearby villages at a local swimming hole. It was here that she encountered a young girl whose story has stayed with her:

> The most profound event was the young 6 to 8-year-old girl that came in with a machete wound from one of the other kids at about 2am one night.
>
> It was difficult as we were not allowed to treat the locals so I could only bandage the little girl's arm. The next morning she started the journey to the local medical centre about half a day's paddle and walk away.[15]

Now at Malu'u, Tracey was co-located with a 6RAR platoon. As an Advanced Medic, she was asked to join a foot patrol, providing security for civilian police who were trying to restore law and order:

> They wanted an advanced medic … two sections were going out for a three-day period. This was one of the most memorable times of my career.
>
> We had a bit of jungle patrolling in thick vegetation that was very boggy and swampy. We also hired a boat, a car, slept in a hut, and had a lot of road marching as well.[16]

---

5 December 2011.

[14] See chapter 6.

[15] Connors, Tracey, extract from private correspondence with Susan Neuhaus dated 5 December 2011.

[16] ibid.

Now and the future: women and medical command

Tracey's ability as a soldier and leader of men continued to be recognised by her unit when she returned home. After further postings with 1HSB, in 2008 she was promoted to the rank of Warrant Officer Class One and appointed Regimental Sergeant Major (RSM) at 1HSB. It was a tribute to her hard work and style of leadership, where she expected others to achieve their best and make sound decisions:

> I think at times we tend to micromanage our soldiers and not give our junior leaders, especially the corporals and sergeants, the ability to make decisions.
>
> I empower the guys here to make decisions for themselves. If they come to me and say, "Can I go home?" I'll go, "I don't know, can you? Have you finished your work? I don't know, you know." Then I point to their rank, and say, "What's that for?"
>
> It's been difficult for me to get people to make decisions for themselves and for their soldiers. I think Defence needs to not be so risk-averse. Just let the guys do what they need to do.[17]

As a female RSM, Tracey feels that she has brought a new approach to the role. Committed to autonomy, she expects high standards from her subordinates. She leads by example, empowering her staff rather than pursuing an authoritarian approach. Now a Captain, posted to the 1 Psych Unit at Randwick Barracks, Tracey continues to stand by the Army's values of courage, teamwork, respect and initiative, seeing gender as no barrier to her career:

> I have found that I have not had to really work harder, but be myself, and in most instances this has served me well. I have natural physical ability that has allowed me to gain the respect of male superiors, peers and subordinates much easier.
>
> Through my career postings, especially at SCMA [Soldiers Career Management Agency], I have gained a substantial amount of knowledge to be able to provide the right advice on the spot to my commanders.

---

[17] Connors, Tracey, extract from interview recorded with Sharon Mascall-Dare on 29 October 2013.

> As a female RSM there is always the misconception from subordinate male members that you are weaker, not only physically, but mentally as well. For me this has never really been a problem due to the fact that I am taller and fitter than a lot of my subordinates. I generally have a lot of knowledge on the regimental side of the military and so it does not take long before the male members are 'in line'.
>
> There is always one or two that try to 'pull the wool over your eyes'. However, it does not take long for them to find their place and move forward. I have been told on numerous occasions that I am intimidating, so I believe this adds to the fact that I have no trouble with disciplining any male members. Generally I am direct and this is met with respect and compliance.
>
> It is said there is a 'glass ceiling' and that women are not as competitive as the males, however I have not found this to be the case.[18]

Brigadier Georgeina Whelan AM CSC

The notion of a 'glass ceiling' is equally foreign to Brigadier Georgeina Whelan, who is set to become the first female Head of Corps of the RAAMC. With the support of her husband and four children 'George' – as she is known to colleagues and friends – has brought about the modernisation of health services and command structures in the ADF.

Born in Dublin, Ireland on 27 December 1965, Georgeina's childhood also had its challenges. The family migrated to Australia in 1972 and as the eldest of five children Georgeina was always aware that money was tight. Knowing that her family would struggle, financially, to send her to university she was unsure of her future. Then one day she saw the Army recruiting van.

> I had applied for public service jobs, jobs in the bank, I had been knocked back by everybody. I was feeling pretty flat when I

---
[18] Connors, Tracey, extract from private correspondence with Susan Neuhaus dated 5 December 2011.

saw the recruiting van and walked in. I said, "I want to join the Army".[19]

There were more setbacks before Georgeina was accepted. She appeared before an Officer Selection Board and was told to come back in 12 months – not an option in her view. Determined to start her career, she enlisted as a soldier. After graduating from Kapooka and a short stint in the Psychology Corps, she applied to join the RAAMC as an officer. This time she was successful and completed her officer training as an exchange student at the Officer Cadet School, New Zealand.

Her first appointment as a General Service Officer in the RAAMC was as an Administrative Officer at the 2nd Military Hospital. Six months into her tenure she was posted as the Adjutant of the newly amalgamated 1st Field Hospital and the 2nd Military Hospital. Those early years, as a medical administrator, became a foundation for her career:

> As an administrator, working alongside wonderful healthcare professionals, I was learning not just what they do, and the importance of what they do, but what my role was as an administrator, in terms of supporting them.
>
> So my early career was in base hospitals, a military hospital, camp hospitals, and then I moved into the field hospital and field ambulance environment.
>
> Over the years, I've had that balance of garrison and combat health experience at all rank levels.[20]

It was experience that proved to be invaluable. Not only did Georgeina have the opportunity to live and work with servicemen and women at all rank levels, she developed an understanding of their needs and concerns. She connected with her soldiers and developed a leadership style that reflected her understanding of their expectations. "Having been a soldier for a couple of years probably gave me a very good appreciation of our workforce and the importance of good leadership," she says. "As a soldier, you're highly critical. You pick up on those solid, strong leadership traits

---

[19] Whelan, Georgeina, extract from interview with Sharon Mascall-Dare on 2 November 2013.
[20] ibid.

and those individuals that you don't want to work for. Who you don't want to be led by."[21]

Her role as an administrator also enabled her to develop rapport while steering clear of politics. As a General Service Officer she observed professional rivalry between doctors and nurses, but was not part of it. She also challenged perceptions regarding the allocation of traditional roles: although clinicians were usually earmarked for command positions, she proved herself to be equally proficient as an administrator. After completing an exchange training post to the United States Army Military Medical Centre and School at Fort Sam Houston, Texas in 1997, she was promoted and posted as the Operations Officer of 1HSB. Later on in her career she would become the Commanding Officer of 1HSB; the first woman and only the second CO not to be a doctor.

If she encountered a 'glass ceiling' early on in her career, it was that administrators had not been considered for such command roles in the RAAMC previously. In the same way that Georgeina refused to be limited by her gender, she worked hard to "break barriers" associated with her role:

> It's hard work. I have to remind myself that because I'm not a clinician, I'm not a doctor, I'm not a nurse, I have to be guided, I have to be inquiring, I have to ask the questions. And there are days there were I don't have the answers purely because of my background and my own education.
>
> There's a number of clinicians out there that, over time, have been quite disappointed with my appointment into certain jobs – I think because I was breaking the barriers, breaking down tradition and challenging the norms.
>
> Over time, some of them did determine that it was an appropriate decision. But others, to this day, are very traditionalist from a military sense. They don't appreciate or support the fact that I *am* an administrator and I *am* in these jobs.[22]

---

[21] ibid.

[22] ibid.

Now and the future: women and medical command

As her career progressed, Georgeina continued to excel as an administrator in positions traditionally held by clinicians. She deployed as the Operations Officer of 1HSB to INTERFET in East Timor in September 1999; she completed the Royal Australian Navy Staff Course in 2000 and was awarded the Lonsdale prize for leadership. As a newly promoted Lieutenant Colonel in 2003 she was posted to Land Headquarters, where she developed and implemented the Land Command Health Continuum with a focus on Mental Health, Injury Prevention and Rehabilitation. In 2004, she was awarded a Conspicuous Service Cross in recognition of this work.

Her appointment as CO of 1HSB in January came at a key moment. On Boxing Day 2004 – the day before Georgeina's 39th birthday – an earthquake beneath the Indian Ocean caused a devastating tsunami, costing more than 300,000 lives. The epicentre was off the coast of Sumatra and the city of Banda Aceh became the centre of a humanitarian crisis. As the world scrambled to help, Australia's response was Operation Sumatra Assist, including a 1HSB contingent commanded by Lieutenant Colonel Whelan.

Georgeina remembers the mobilisation phase as both organised and chaotic. Information was sketchy at first and as the hours went by the true scale of the disaster began to emerge. At first, Georgeina selected a captain to head the Army's 10-person resuscitation and environmental team, but two days later the medical capability had grown into a 113-person team. Specialist medical reservists were high on the list of those asked to deploy – they had the required capability and many were already on Christmas leave, although many had to request additional time off and arrange cover for their civilian practices in order to complete the five-week deployment.[23]

As news of the tsunami's impact continued to feed through, Georgeina did what she could to pack enough equipment. Ten beds became 200 beds, as pressure mounted:

> I got a phone call saying, "How many beds have you packed?"
> And I said, "I've packed the amount of beds you've told me to pack." And they said, "No, no, no. We've promised the Prime

---
[23] See: Jamieson, Cameron, 'Commanding Decisions', *Defence* magazine, 1 March 2005.

> Minister 200 beds. You need to take 200 beds."
>
> I shouted, "Have you got any spare stretchers out the back? Chuck those in as well!"
>
> We literally packed everything but the kitchen sink and got out to the airfield.
>
> At that stage, even my Brigade Commander was saying, "We're still not quite sure where you're going. Maybe by the time you get to Darwin for refuelling, we can tell you where you're going."[24]

Her destination was to be Banda Aceh. Georgeina's first task was to identify a suitable location for a field hospital – a place that would reach those affected by the disaster and leverage Australia's medical capability as part of a multi-national relief effort. Georgeina's RSM identified the site of a local teaching hospital – Zainal Abidin Hospital – that had been devastated by the tsunami, but was known to local people. With help from the PST – who began by establishing a light surgical facility in front of the hospital – and colleagues from a German military hospital who had also arrived, ANZAC Field Hospital began treating casualties. Before they could treat casualties however, there were mountains of debris and mud that needed to be moved, and the hospital had to be scrubbed and cleaned to make it suitable for patients.

At times the scene was confronting. During her reconnaissance of the area by helicopter, Georegina was struck by the numbers of bloated bodies strewn among the broken trees:

> I had 20 years in the medical corps by this stage. I had seen death in East Timor as the operations officer, but there was always a degree of distance as the 'OPSO' from what the clinicians did. You're also in a group and you're all experiencing it together.
>
> I remember there were times when I was emptying out the 150-cubic-foot fridge because we had no morgue to put the bodies in. But it was all very clinical, very organised and it didn't really impact me too much.

---

[24] Whelan, Georgeina, extract from interview with Sharon Mascall-Dare on 2 November 2013.

Now and the future: women and medical command

Lieutenant Georgeina Whelan in Banda Aceh in the immediate aftermath of the earthquake.
Image: Department of Defence

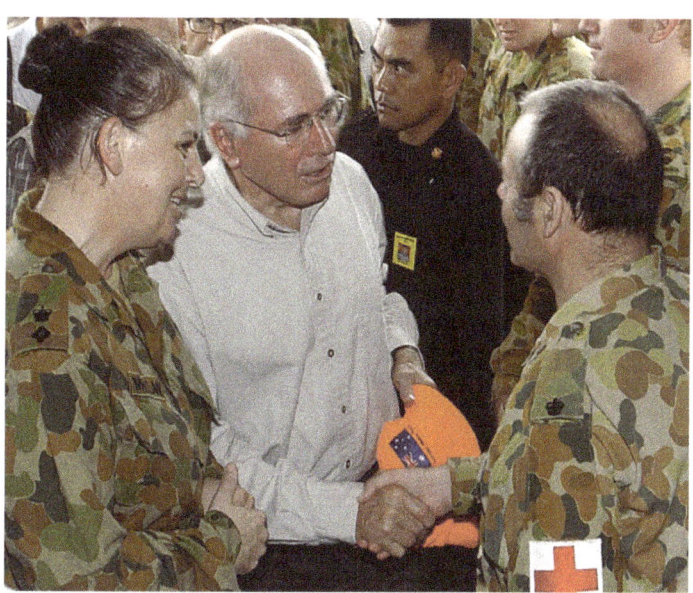

The Australian Prime Minister, John Howard, is introduced by ANZAC Hospital commanding officer Lieutenant Colonel Georgeina Whelan to Chaplain Don Kaus, of the Australian Army 1st Health Support Battalion, at the hospital in Banda Aceh during Operation Sumatra Assist.
Image: Department of Defence

> But when I first arrived in Aceh, my RSM and I were flown up in a helicopter from Medan to do the reconnaissance. We were in a vehicle and there were just bloated dead bodies floating.
>
> I wasn't ready for it. I was really confronted by it.[25]

Despite her initial reaction, Georgeina stayed on task. Looking back, she acknowledges that the scene had an impact, but as the CO she had a job to do. She found a way to "box away" her thoughts and get the hospital up and running. "It was really hard for us to do our job initially in Aceh. We had to fight to get in and fight to do our job. Once people realised who we were and what we could do, the rest is history."[26]

The story of what was achieved at ANZAC Field Hospital did indeed make history. As Georgeina and her team moved through the stages of the operation, from engaging local truck drivers to provide transport, to managing the outbreak of disease in the aftermath, to reconstruction and sustainment of health facilities destroyed by the tsunami, they worked hard to explain their work to the media and secure public support back home and on the ground. From the outset, the Army knew it had to communicate its role clearly and media training was an important part of Georgeina's training before she left. Thrust into the spotlight on local and international media, Georgeina gave a human face to the ADF's contribution: as a woman, she conveyed both compassion and competent professionalism.

> I think people saw the human side of the ADF and got to know the personalities of the individuals working in Aceh – whether it was the medics or the doctors, whether it was the engineers or the aviators.
>
> The government was very supportive of what we were doing; the ADF was very supportive; our community was very supportive.
>
> It can be really hard in the ADF to do your job and send the right message back to Australia. It was many years later that

---

[25] ibid.

[26] ibid.

Now and the future: women and medical command

> I started to realise, as I was growing and maturing, that there were implications politically. I realised what we had done in terms of contributing to better relations between Australia and Indonesia.[27]

Her deployment to Sumatra had changed Georgeina as an individual. She returned to Australia with a far greater awareness of the correlation between command structures and capability – her experience motivated her to push for improvements in training for RAAMC staff and changes to service delivery. In 2007 she was posted as the Senior Health Officer of the Area Health Service, NSW where she amalgamated the three NSW Area Health Services into one organisation. By early 2009 she was promoted to the rank of Colonel and posted as the Command Health Officer, Headquarters Forces Command; again the first time a woman, or someone other than a doctor, had held the role. Here she would play a central role in shaping the future of health capability in the RAAMC and the wider ADF.

Over the past five years, Georgeina has realigned the Army's Combat Health Support structures. A new 1CHSB (1st Combat Health Support Battalion), drawn from 1HSB, consists of integral and close health support personnel centralised under Close Support Health Companies (CHSC) located in Roberston, Lavarack and Gallipoli Barracks. 2GHSB (2nd General Health Support Battalion) provides general health support, centralising capabilities such as pathology, imaging and physiotherapy. 3HTRB (3rd Health Training Reserve Battalion) draws on specialist reservists to provide 'high end' capability. Together, the new units support the Army's Land Based Trauma System, incorporating Force Health Protection and Occupational and Environmental Health. The aim is to provide timely, appropriate healthcare and support both during training and on deployment. It was a restructure required to meet the demands of military healthcare in the 21st century:

> I was not convinced we were structured appropriately to develop the next generation of health capability. I felt we were not maintaining parity with our allies and in terms of the more

---

[27] ibid.

contemporary methods of generating deployable health capability and then delivering that across the battle space.

It's still early days yet, but the Chief of Army said to me recently, "It's not a popular decision that I've made based on your recommendation to follow through on the Combat Health restructure, but it was a very necessary decision because it's enabled us to salvage and continue to regenerate a capability."

There's still a long way to go with it, but the intent is to generate a capability that can live, work and operate alongside the soldiers, sailors and airmen it deploys with.

It's a change of command status so that health can drive its own agenda to focus on training, management, and professional development. Issues that have taken a backseat to the other training continuums that must be conducted in Army.[28]

In 2014, the now Brigadier Georgeina Whelan can rightfully claim her position as leader and innovator in the Australian Defence Force. She has prioritised training for RAAMC personnel to ensure they are ready to deploy on par with their overseas allies, and do not merely support the pre-deployment training of other corps. She has positioned health as a core capability. Under her leadership the RAAMC continues to strive for professional excellence, from an historic foundation. From surgery to pathology, from imaging to physiotherapy, members of the RAAMC are specialists in what they do.

As a wife and mother, daughter and sister, Georgeina also embodies the evolution of women's roles in the Corps. Like many of women who have served before her, she has faced the guilt of leaving her children to go away on deployment. She has relied on her husband to run the household in a society that still regards stay-at-home fathers as an aberration. Today's women of the RAAMC expect an adaptable, flexible working environment and the challenges they face are different from their male peers.

The professional landscape for women of the RAAMC has also changed. The impact of specialisation of the medical workforce, across all areas of healthcare has also seen 'general' skills eroded, in medicine, radiology,

---

[28] ibid.

pathology, physiotherapy and scientific roles. There has been an increasing emphasis on healthcare administration and the development of non-clinical command streams. In the future it is likely that there will be much greater 'porosity' between the Regular Army and the Reserve component, with a greater emphasis on civilian integration and professional military mastery.

The medical women of today's Army are required not only to be technically proficient in their area of healthcare, but they are required to have a mastery of military skill unknown to their forebears; they are physically fit, wear combat body armour, are competent at firing assault rifles and personal weapons and are expected to have an understanding of the politics and tactics of military operations.

In 2011, the first woman (Navy Rear Admiral Robyn Walker AO) was appointed Commander Joint Health and Surgeon-General of the Australian Defence Forces. In 2014, Brigadier Georgeina Whelan is set to become the next Head of Corps of the RAAMC. Through its appointment of women, the ADF is embracing a new generation of leaders who bring with them new priorities, new ways of working and a new approach to the future.

Epilogue: reflections on service

Susan Neuhaus CSC, former Colonel RAAMC

This is not my story, but instead the stories of many women whose lives have inexorably meshed, one way or another, with my own: linked through a centenary of professional women and their commitment to this country.

In March 2009 I was serving as a Colonel in the Australian Army posted to Afghanistan as a surgeon. Like the other 3,000 troops on base, mostly Dutch and Australians, this was my temporary home – a Spartan landscape of shipping containers and Hesco barriers. Morning walks were punctuated by the occasional 'G'day' as I passed unfamiliar faces in Australian combat fatigues (usually sporting several months of facial hair), and eddies of thin, icy air whipping between the rows of shipping containers, piercing layers of thermal clothing.

One morning, not unlike any other, I sat on a hard metal chair in the medical facility, the icy cold penetrating the steel floor and the reinforced steel walls. For me, it was not so much as if I was in the mountains of Afghanistan; rather that I existed in a bubble, where from time to time a piece of Afghanistan would enter into my world. On this morning, the father of a young Afghan boy sat in front of me, his face weary and suspicious. In his case though, it was not the war that had brought him here, but an avalanche: a treacherous mountain slope that disintegrated under the weight of winter, crushing his son under a mass of rock and ice.

We spoke through an interpreter – seeing each other but unable to fully digest each other's words. I found myself wondering what he was thinking – torn from his village with its ancient customs and transported to this

Epilogue: reflections on service

First-World medical facility in the mountain slopes of Uruzgan. It must have seemed unreal, even mythical to this man, to be immersed in such an Aladdin's Cave of technology. Despite our best efforts to communicate, there was also a cultural chasm. What on earth did he think ... of women in the same uniform as men? ... of their comfortable equality? ... of a female surgeon? ... a woman in a leadership role? Regardless of his thoughts, he trusted us with future of his boy, his only son.

Like most armies, we were not equipped for paediatric work. We would try to save his son's arm, I said, but I knew it was unlikely. It is a tragedy for any young boy to lose a limb, but in this society – destroyed by war and with inadequate medical facilities – the repercussions would be greater. Loss of his right arm would see him excluded and ostracised from his friends and peers – yet this boy had the love of his father. He was among the lucky ones.

We saw many children here; some the victims of conflict, but most victims of a society torn apart; diseases unrecognised in Australia, but the reality of nations at war – late diagnosis, poor nutrition, and conditions that could be prevented by good water supplies, vaccinations and public health.

For a moment, as I looked into the young boy's eyes, I saw the faces of my own children, thousands of miles away, tucked up in their beds in suburban Adelaide. But for me such memories were 'out of bounds'. I had made that resolution on day one.

In the weeks that followed, my thoughts returned many times to this dissonance; the place of women in Afghanistan's patriarchal society and the place of women in my own military.

Whatever issues these men may have had about women in uniform or in leadership roles, they realised that we were there to help. They lived amongst us in the medical facility, they prayed in our corridors and spent nights of anxious vigil at their sons' bedsides. They saw us at our best and at our worst. And if things did not go well, and they often did not, it was the fathers of these boys who provided us with a sense of absolution.

One of the most humbling and extraordinary experiences of my professional career was to have a father come back some five days after the death of his young (and only) son and, rather than to offer retribution, to thank me, and the team for our efforts: "*Insh Allah*" – "This was God's will" and it was not a failure on our part. I cannot underestimate how far that goes to mitigate any personal sense of failure; nor could I but wonder if it would have been different had I not been a woman.

Several years earlier, Professor John Pearn, an eminent historian, retired Major General and previous Surgeon-General of the ADF, had encouraged me to explore the history of women in the RAAMC. Now I was resolved to search for their stories on my return; to better understand women's roles in the Army and the legacy of women who have served as surgeons, doctors, administrators and other medical professionals throughout the past century. I wanted to discover what their stories meant to me as an Army officer and surgeon, and as a wife and mother, and whether women did in fact bring a different perspective and dimension to the art of military medicine.

I naively thought this would be a small task. After all, I assumed, there could not be that many female doctors who had served with the Australian Army. I knew most of them by name and, although there were female surgeons in the RAAF, I was the only female surgeon amongst my Army colleagues at the time.

Instead, I found myself on a journey of discovery that would extend over five years of research, involving days and weeks spent trawling through dusty archive boxes, photographs faded by time, and sometimes indecipherable pages of handwritten diary entries from generations past. It was a journey that took me to libraries and archives in Melbourne, Canberra and New Zealand. It took me to the Medical Corps Museum outside of London and sites on the Western Front, where battlefield hospitals had cared for the wounded in World War I.

During this journey I discovered more and more extraordinary Australian women. Their pioneering work dated back more than 100 years, from Egypt, France and the mountains of Serbia in World War I to the North African desert in World War II. As a doctor in uniform, I became aware that these women had quietly paved the way for my own career – many fought prejudice and discrimination along the way. The stories of contemporary women in this book reflect changes that I, too, have witnessed – changes in society and military culture.

Epilogue: reflections on service

I joined the Army in 1987 as a medical undergraduate officer. Over the next two decades I had the good fortune to work with both the Regular Army and the Reserve, and serve in a number of roles; as a staff officer, a commander and a clinician. Over those years I have seen many of these changes reflected in the healthcare environment, both at home, and on overseas operations.

Unlike many of the women in this book, my ancestry is not grounded in the military, although my father served briefly with the Citizen Military Forces (CMF). My earliest military memories are those of Anzac Day as a child, being awoken from my nice warm bed, rugged up and taken out by my father in the dark to a chilly pre-dawn service that I did not really understand. I have memories of the silence, half-murmured hymns and very old men – as they seemed to me then – gathered in their own thoughts and reflections. They would always stand tall and proudly to the strains of "God Save the Queen".

Later in life, after I had enlisted in the Australian Army and had deployed overseas, Anzac Day took on a different meaning for me.

Anzac Day in 1993, during my deployment to Cambodia, remains seared in my memory. I don't recall the traditional rum and coffee, but I do recall the rising heat of the morning and the strains of the bugler competing with the local music blaring from village speakers strung up on trees next to our compound, and the sound of pigs and chickens running amok in the streets. I vividly remember the New Zealand contingent, many of whom worked with the mine-clearance teams, performing the first Haka I had ever seen. It was extraordinary to stand there that morning with the Regiment, the first major deployment since the Vietnam War, and be aware that each of us were a small part in the thread that links Anzac Cove to the present.

A few months later I would be reminded, again, of the loss and grief that is intrinsic to war when, in 1993, another female doctor – Major Susan Felsche – was killed in the Western Sahara: the first Australian woman to be killed on operations since World War II. Susan's death was testament to the inherent risks of peacekeeping operations – whether they are from conflict or the environment – such missions are not benign.

On Anzac Day the following year, back in Australia, I marched for the

Colonel Susan Neuhaus (right) in Kuwait, after barrier weapons testing of the range required before entering Afghanistan (with Major Ken Wishaw, consultant anaesthetist and intestivist).
Image courtesy of Ken Wishaw

first time in uniform. In those early days of the Peacekeeping era, very few serving officers had medals, and women with medals were even fewer. I recall the stinging looks from those who questioned my operational service – those who thought I was too young, or those who had served loyally for 20 or 30 years without having the same opportunity to deploy.

For many years to come, on the Anzac Days that followed, well-intentioned veterans would tell me my medals were on 'the wrong side'. More recently, after returning from Afghanistan, I was approached by someone who on noticing me with my two daughters at a Dawn Service expressed concern that I was a 'war widow'. It was a question that took me by surprise, and a poignant reminder that serving women and female veterans remain largely unnoticed and unacknowledged in our society.

There is no doubt that the Australian Defence Force has been through an era of unprecedented operational activity with repercussions for women's roles and how they are perceived. After the 'long peace' that followed the Vietnam War, Australia has been involved in peace-monitoring, peace-enforcement and disaster-relief operations in Africa, the Middle East and

Epilogue: reflections on service

the Asia-Pacific region as well as combat operations in Iraq and Afghanistan – and women have taken roles in every one of these operations. With this has come new opportunity and diversity, an expansion of roles for women and a drive for the pursuit of equality.

As I transited into Afghanistan in early 2009, the officer commanding the Kandahar-based Australian helicopter squadron was a woman. Army has now opened all roles to women, and from late 2016 women will be directly recruited into frontline combat positions. While this new development continues to be the subject of public debate and media scrutiny, in researching this book I have been reminded this is not new. Women have been exposed to the dangers of frontline service since World War I. Quietly, they have got on with the job – continuing today much as they did in 1915.

As we approach the Anzac Centenary next year, the stories of World War I that have shaped our national identity will be forefront. We will remember the fractured families, and the devastated communities bearing name after name from the same family on their war memorials. We will remember a conflict that killed 60,000 Australians killed and left another 156,000 wounded, gassed or taken prisoner.

But as we remember, we should not forget the extraordinary women who also did their part both in World War I and conflicts since. This book is but one contribution to their stories, with its focus on women who served as medical professionals with the Australian Army and its allies. It was never meant to be an 'official history' – and there will no doubt be omissions. Instead, it is a collection of stories that contribute a far larger narrative. It tells the stories of Australian women who, often despite prejudice and inequality, donned khaki uniforms and served their nation with as much distinction and valour as their male counterparts in all of our conflicts from World War I to the current day.

I did not know these stories until the end of my military career. I wish I had known of them at the beginning, because here, amongst these pages, are the missing female role models; pioneers, leaders and women of indomitable sprit.

Notably, many of the women in this book have not been recognised for their service. Others have been, but with lesser awards than their male colleagues. Major Josephine Mackerras' story in particular reminds us that despite professional excellence, women continued to experience discrimination through their exclusion and omission from honours and awards. But these women did not seek glory. They have simply and quietly 'got on with the job'; a 'job' that has sometimes been conducted under extraordinarily arduous and dangerous conditions. The women in these pages deserve to be acknowledged. With their quiet strength and determination they are role models for the current generation. Their stories are important – and in coming to know them, and sharing their journeys, I have gained valuable insights into my own.

The story of Captain Vera Scantlebury is a case in point. Vera was 28 when she travelled to London and accepted a position at Endell Street – a year older than I was in Cambodia. Despite our differences, I can relate to her experiences as a young woman thousands of miles from home, feeling professionally and personally isolated and struggling to deal with the reality of her environment. Like Vera, I discovered that little learned in medical school would have helped. In the early days of the war, Vera spent hours poring over anatomy books, in the same way that I spent sleepless nights going through tropical-disease textbooks in Cambodia, attempting to master the mysteries of obstetrics in Bougainville, or learning how to deal with the devastating effects of a suicide bomber in Afghanistan and the harsh aftermath of triage and mass casualties that no amount of training or practice could ever prepare for.

Vera's personal circumstances also struck a chord. She clearly found it difficult to choose between her profession and her aspiration to be a dutiful fiancée. Like other women in this book I have known love and separation. In Cambodia, I was young and in love, although not yet engaged; in Bougainville I was married. In Afghanistan, I was the mother of two small girls I had left at home. Like most young doctors, my personal commitment to Army was sorely tested at times. I felt that professionally I was lagging behind my civilian peers while I was under a return-of-service obligation. But nothing could have been further from the truth. The experiences that Army offers are things that civilian counterparts could only dream of.

I can also appreciate Vera's difficulty in coming to terms with the military

Epilogue: reflections on service

protocols. To an outsider, the Army can appear mysterious – joining up is like joining a tribe, with each Corps having its own distinctive (and not always understandable) rules, codes and traditions. I was fortunate to start my career in the regimental environment of Kapooka where I was infused with the fundamentals of basic soldiering and the functions of a platoon commander. I came to understand the relationship between doctor and commander, how to thrive in a military culture and how to navigate the complexity of competing obligations and responsibilities.

Similarly, as I researched Agnes Bennett's story I sensed her frustration and harshness of the environment – the strain shows in her diaries. Battlefield medicine – whether in World War I or modern-day Afghanistan – is characterised by uniquely complex medical and ethical issues. The impact of multiple, severe mutilating trauma casualties, infant casualties, end-of-life decisions, fatigue, adverse medical outcomes, varying cultural values and the evidence of inhumanity can take a large personal and collective toll. Still, Agnes's leadership skills were exceptional in keeping her team together. She was a woman unafraid to make tough decisions, even at cost to herself.

As a surgeon, I can identify with Lilian Violet Cooper's sense of professional fulfilment as she worked tirelessly during the Serbian campaign, but the conditions under which she worked through that bitter winter were far more austere and far more difficult than any I could imagine. Dobraveni dressing station is in no way comparable to the sophisticated ballistic protection of my operating theatre in Afghanistan with the most modern available medical equipment and technology – for example, frozen blood products not yet even available in Australia. Nor is it comparable to the tented operating theatre in Bougainville; with telephone access to specialist advice, helicopter retrieval and on site pathology and imaging. And yet Lilian and the other women did not just prevail under these circumstances, their clinical outcomes could match those of any sophisticated military force today; testament not only to their endurance, but to their courage, skill and judgement.

Women are no longer required to leave Army service when they marry or have children. Nonetheless, mothers deploying on operations is a recent phenomenon in our military; or so I thought. Majors Mary

Thornton and Josephine Mackerras who served in World War II disabuse this fiction. Both were outstanding professionals in their own right, and both challenged convention by pursuing military careers while having children. Mary's decision to leave her son in boarding school to pursue her commitment to the war effort was clearly driven by a passion for service. Her decision to return to the war in the Middle East after the death of her husband is however, by any measures extraordinary. We cannot know the sacrifice to both mother and child.

For me, the era that followed the Vietnam War has particular resonance, given the wealth of stories involving women I have known personally or professionally. To tell these stories, I have drawn on the interviewing skills of my co-author, Sharon Mascall-Dare, and her expertise as a former BBC journalist and documentary producer. Sharon has brought a new perspective to contemporary stories that, in some cases, I knew only too well, but in other cases was oblivious to.

My first posting was to Kapooka, the Army's 'School of Soldiering'. It was here that the Adjutant, a young captain called Georgeina Whelan, taught me how to wear my uniform, how to march and how to salute. It was invaluable instruction that gave me a foundation for my new career. Georgeina's journey has been extraordinary and she is one of many women in this book who I deeply respect and admire. She is also one of few who are not clinicians, but a woman who has been able to navigate an often tenuous pathway between the clinical and non-clinical world and carve a new pathway. As she prepares to take on the role of RAAMC Head of Corps, it will set a new agenda for the Corps, and not only in terms of gender.

The stories in Chapter 6 (Other People's Wars) have a very personal connection. Sergeant Norma Hinchcliffe and Corporal Liz Matthews both worked for me when I was the Regimental Medical Officer for the second UNTAC rotation in 1993 – a year that in many ways set the foundation of my military career. The challenges that confronted me there were those that confront all junior leaders: building the team, managing the diversity of skills and experience – along with the usual strains of providing medical support in a difficult and complex environment. Broad clinical responsibility was mixed with new and exotic medical conditions to treat, the realities of being on duty 24/7 and the inherent challenges of working and living among your patient population.

The experiences of the female medics represented here typify those of all the soldiers I have worked with in these environments and reflect the evolution of medic roles from 'medical orderlies' to highly proficient and skilled individuals who work across broad domains of healthcare: including as emergency first responders, in key trades such as pathology and radiology as well as traditional clinical environments. I continue to be amazed by what Australian soldiers can do – and what I, in turn, have learned from them. Australian soldiers are renowned for their ingenuity and I have seen our medics (male and female) do amazing things – from constructing a humidicrib from nothing more than fruit boxes and fluoro lights in Bougainville, to delivering babies or demonstrating extraordinary compassion when confronted by a dying villager in Afghanistan. In many ways, it is the soldiers who have been the best teachers of egalitarianism, of a 'fair go' and of universal compassion.

Captain Tam Tran and I served briefly together in the same unit. Whilst I knew a little of her background, it was not until we corresponded about this project that I came to appreciate the depth of her experience as a refugee and how that impacted on her service, and I became aware of the experiences 'new Australians' bring to the modern ADF. As our society becomes more multicultural, the ADF is drawing on a wealth of experience and cultural diversity in its ranks. Captain Tam Tran is the archetypal success story – a survivor of war who chose to serve the country that welcomed her. She not only a role model to her community – she is a role model to our society.

The stories of Lieutenant Colonel Su Winter and Corporal Jacqui de Gelder typify the realities faced by women in the modern ADF, as it seeks to become more adaptive to women's circumstances and needs. Su is both a medical professional and an 'Army wife'. She knows, first-hand, the reality of trying to accommodate two military careers in one household. Jacqui's depiction as a front-line soldier – emblazoned in the media – is a reminder that women's roles in the military remain a subject of debate and contention. But despite the views of the popular press, Jacqui's experiences of the military echo my own: that at its core the Australian Army is a meritocracy — and division lines are not so much about gender, but about role and about performance. Like Jacqui, I have been both accepted and exceptionally supported by the men I have worked with and for, even in situations that might otherwise be thought problematic.

The final chapter of this book is testament to how far we have come. Far from being told to 'go home and knit' as Agnes Bennett was instructed in World War I, Warrant Officer Class One Tracey Connors has earnt her 'wings' as a member of the elite Parachute Surgical Team and her 'crossed rifles' as an accomplished markswoman. She is a leader of men and women and an inspiration to many. Like other women who have also reached the rank of Regimental Sergeant Major (RSM), she has demonstrated that attitudes to women's career progression and recognition are changing.

The appointment of Brigadier Georgeina Whelan as RAAMC Head of Corps is a culmination of such change. Unthinkable 100 years ago, when women doctors were excluded from joining Australia's war effort, her achievement stands for even greater commitment to equality and diversity in the modern ADF. Married, with four children, Georgeina's career challenges many of the prejudices and assumptions of the past. As we approach the Anzac Centenary, she will lead the RAAMC into a new century of service for Australia's military medical professionals — both women and men.

In 1919, Winston Churchill, then Secretary of State for War, said:

> The command of Medical Field Units involves leadership and discipline, and at times very great strain and hardship to which women would only be equal in very rare cases.[1]

As the stories of the women in this book demonstrate, Mr Churchill was wrong.

In the dedication to this book, both Sharon and I acknowledge our daughters — Grace, Emma and Claudia. In passing on these stories, we hope to inspire the next generation of young Australian women as we prepare to commemorate a century of service and the Anzac Centenary.

---

[1] The War Office, document dated 2 May 1919 in CMAC:SA/MWF/C.163, quoted in Leneman, Leah, 'Medical 'Women at War 1914–1918', *Medical History*, 1994, 38: p. 174.

Epilogue: reflections on service

Colonel Susan Neuhaus CSC with her daughters, Pulteney Grammar School Remembrance Day Service, 2011.

Whilst the world our daughters will inherit will be no doubt bring new challenges, it is important that they know and understand the stories of women who have gone before; not just of what Australian women are capable of, but what Australian women have already done.

It has been an immense privilege to have uncovered these stories. It is a greater privilege to be able to share them.

Appendix 1: Chronology of medical women in the Australian Army

The 1900s to the 1940s

1901   Army Nurses serve in the Boer War on the establishment of the Australian Army.

1902   Australian Parliament passed the Commonwealth Franchise Act 1902 enabling women to vote and stand for election to the Federal Parliament.

1903   Formation of the AAMC.

1914   Between 1914 and 1919 members of the Australian Army Nursing Service (AANS) serve overseas. Alongside them are female masseuses (physiotherapists). Both nurses and masseuses are entitled to the privileges of commissioned rank, but do not, formally, hold any rank.

1918   Phoebe Chapple becomes first Australian doctor to be awarded the Military Medal.

1939   Nurses and physiotherapists serve overseas as part of the 2nd Australian Imperial Force (AIF).

1940   The first female Medical Officer is appointed in the AAMC (Lady Major Winifred MacKenzie).

1941   A shortage of males leads to the formation of Women's Services in the 1940s. There were over 60,000 women in the three Services.

Appendix 1: Chronology of medical women in the Australian Army

1942 Formation of the Australian Army Medical Women's Service (AAMWS), which originated during WWI in the Voluntary Aid Detachments raised from the Australian Red Cross and the Order of St John.

1944 War Financial Regulations amended to allow all prospective women officers of the AAMC, irrespective of their specific health discipline, to be paid at the same rate as female officers in the other services (AANS, AWAS and AAMWS). Their pay remained less than male officers for the same rank, qualification and posting.

1948 Royal Prefix is granted – Royal Australian Army Medical Corps.

The 1950s to the 1960s

1950 Government approves in principle the re-formation of the Women's Services – Women's Royal Australian Air Force (WRAAF), the Women's Australian Army Corps and the reconstitution of the Women's Royal Australian Naval Service (WRANS).

1951 Enlistment of women into the Women's Services does not exceed 4 per cent of the mainstream services. Women remain managed by women and non-trade training is conducted by women. AAMWS disbanded.

1967 Women officers in the RAAMC serve in Vietnam.

1969 Women are allowed to remain in service following marriage.

The 1970s

1974 Pregnancy does not automatically mean discharge.

1975 International Year of Women. Chiefs of Staff Committee agree to set up a Working Party to examine and report on the role of women in the Australian Defence Force (ADF). It is recommended that women should be permitted to serve on active service, but not in combat roles.

1979 Women receive equal pay to males. Women's Royal Australian Army Corps (WRAAC) is abolished and amalgamated with mainstream Army.

The 1980s

1980   The Australian Government ratifies the International Convention on Elimination of All Forms of Discrimination against Women. Defence requests exemptions for combat and combat-related duties.

1984   Sex Discrimination Act promulgated. The ADF gains an exemption allowing it to discriminate against women in positions involving the performance of combat duties or combat-related duties; or in prescribed circumstances in relation to combat or combat-related duties. 17,000 ADF positions (23.5 per cent) opened to women in competition with males.

1986   A total of 21,750 ADF positions (35 per cent) opened to women in competition with males.

1987   First two female pilots graduate from the RAAF pilots course.

1989   A total of 28,562 ADF positions (43 per cent) open to women in competition with males.

The 1990s

1990   Army sets up the Combat Related Employment of Women Evaluation Team (CREWET).

1992   'Halfway to Equal' (the Report of the Inquiry into Equal Opportunity and Equal Status for Women in Australia) recommends that Defence exemptions under the Sex Discrimination Act be rejected. Some exemptions are retained.

1993   Major Susan Felsche is the first woman to die on overseas operations since World War II.

1996   Captain Carol Vaughan-Evans is the first woman to be awarded the Medal for Gallantry.

The 2000s

2012   On 27 September, the Australian Government formally agrees to the removal of gender restrictions from Australian Defence Force combat roles.

Defence Minister Stephen Smith announces that all 21 recommendations of the Broderick Review into Treatment

Appendix 1: Chronology of medical women in the Australian Army

of Women in the ADF have been accepted by the Australian Government. Women are recognised as essential to the sustainability and operational effectiveness of the ADF; recruitment and retention of women is prioritised.

Appendix 2: Harkness Memorial Medal

This award is designed as a tribute to the service given to the Corps by the late Geoffrey Harkness OBE, ED, who served with the Corps from December 1941 to May 1971. Colonel Harkness served as a Regimental Medical Officer in New Guinea and as a Field Ambulance Officer and played a significant role in the development of the Corps. His loyalty to the Corps and unselfish dedication were a hallmark of his service.

The award is funded from a special fund raised by subscriptions from senior officers of the RAAMC and members of the civilian medical profession.

The award is given for contribution to the RAAMC. This must be in every way of an outstanding nature. Enthusiasm, selflessness, dedication and devotion to the Corps are necessary contributory factors. The recipient must be a serving officer of the RAAMC.

This prestigious medal has been awarded annually since 1971. Female recipients of the award are:

2004	Lieutenant Colonel Georgeina Whelan CSC
2011	Major Elizabeth Barnett
2012	Captain Stacey Austin
2013	Captain Kelly Dunne

Appendix 2: Harkness Memorial Medal and C. F. Marks Award

C. F. Marks Award

This award is designed as a memorial to the late Colonel Charles Ferdinand Marks OBE, ED, who had a long and distinguished career in the RAAMC. He served in the Middle East and New Guinea during World War II. He was subsequently Commanding Officer of 7 Field Ambulance and 11 Field Ambulance and appointed to Colonel in 1955.

The award is funded from a donation generously given by his widow, the late Mrs J. Marks, and his daughters, who have carried on the tradition.

The award is given for an outstanding individual effort where the RAAMC has benefitted in some way. The emphasis is on contribution to the Corps. Enthusiasm, selflessness and dedication to the Corps are necessary contributory factors. The recipient must be a serving Warrant Officer or Noncommissioned Officer of the RAAMC.

Female recipients of the award are:

2001	Sergeant K. Pullen
2004	Sergeant Virginia Morris
2005	Warrant Officer Class Two Susanne Gibbs
2011	Warrant Officer Class Two K. McNaught

Index

A

1 Australian Field Hospital 156
2 AWH (Australian Women's Hospital) 128, 144
AAMWS (Australian Army Medical Women's Service) 75, 130, 134, 135-41, 313
Afghanistan 152, 185, 227, 253, 258, 259, 266, 271-80, 300-310
America Unit No. 1 48-9, 58
Anderson, Dr Elizabeth Garrett 6, 13, 21-5, 70, 26, 29, 30
Andersons, Colonel Vikija 227-9
ANGAU (Australian and New Guinea Administrative Unit) 143
Ardill(-Brice), Dr Katie 64-5, 71-2
Army Massage Service, *see also* Physiotherapy, 66
Armstrong, Millicent 72-3
AUSMTF (Australian Medical Treatment Force) 279

B

Bali 231, 252-8
Banda Aceh 293-7
Bangka Island 118, 128
Barry, Dr James 1-6, 7
BCOF (British Commonwealth Occupation Force) 130
Bean, C. E. W. 68
Best, Lieutenant-Colonel Kathleen 140
Bedford, Miss Josephine 35, 44, 48, 51, 52, 54-7, 71
Bennett, Dr (Captain) Agnes 18, 20, 34, 44-51, 52, 53, 54, 56, 58, 65, 72, 310
Blackwell, Dr Elizabeth 8
Bougainville 128, 140, 173, 185, 216-25, 227, 240, 254, 306, 307, 309
Bourne, Dr Eleanor 20, 29, 63
Brown, Lieutenant Hannah 265-7
Buckley, Dr Emma 20, 29
Bulkley, Margaret 3, 4
Bullwinkel, Sister Vivian 118, 128
Burns 67, 113, 114, 121, 157, 163, 168, 193, 253, 254, 256, 257, 272

Index

C

Cambodia 171, 173, 175, 174-88, 197, 303, 306
Cambridge Military Hospital 60
Champion, Dr Rachel 20
Chapple, Dr (Major) Phoebe 20, 59-63, 312
Choubra 47
Churchill, Winston 18, 59, 310
Cody, Glynneath 136-8
Coffey, Dr Rosemary 168
Coghlan, Major Shirley 155, 167-8
Comnena, Anna 7
Connors, Warrant Officer Class One Tracey 281-90, 310
Cosgrove, General Peter 226, 260, 263
Croix de Guerre 32, 37, 72, 122
Crosier, Captain Joan 144-6

D

Dalyell, Dr (Captain) Elsie 37-44, 65, 70-1
de Garis, Dr Mary 56
de Gelder, Corporal Jacqui 271-8, 309-10
Dietetics 131-4
Dobraveni Dressing Station 54-6, 307

E

East Timor 173, 185, 200-3, 225-43, 259-60, 264, 287, 294
Empire Star 117-8

Endell Street Military Hospital 21-30, 32, 70, 306
Étaples (Anglo-Belge Clearing Military Hospital) 65
Evans (-Neuhaus), Captain (later Colonel) Susan 175, 180, 187-8, 218, 278-80, 300-11

F

Franklin, Miles 44, 48, 57
Foster, Dr Laura 31
Featherston, General R. G. H. 16, 20
Felmingham, Sergeant (later Warrant Officer) Kim 199-203
Fairley, Sir Neil Hamilton 95, 97-8
Fenner, Professor Frank 99
Fortune, Dr Cyril 102-3
Fleming, Staff Sergeant D. 126
Fleming, Captain Gwen 146-7
Felsche, Major Susan 189-98, 303, 314

G

Gas gangrene 22, 37-42, 47, 49-50
Gittoes, George 203-4
Grahame, Major Barbara 168

H

1 HSB (Health Support Battalion) 285, 288, 289, 292-3, 297
3 HSB (Health Support Battalion) 227, 259, 265
Hamilton-Browne, Dr Elizabeth 20-1

Heidelberg Military Hospital 92, 133, 135
Henderson, Private (later Sergeant) Vashti 259-65
Heysen, Nora 100-1
Hinchcliffe, Sergeant Norma 180-2, 308
Holland Park Military Hospital 108, 132, 133
Hollywood Military Hospital 106, 132, 138
Hope, Dr Laura 35
Hôtel Claridge (Hospital Anglo-Belge) 21-23
Howse, Sir Neville 65

I

Inglis, Dr Elsie 34, 45, 46, 47, 48
Institut Pasteur 37, 39, 40
INTERFET (International Force for East Timor) 226, 293
Iraq 90, 227, 244-52, 259-71, 305

K

Kahan, Lieutenant Jean 103-8
Kapooka 176, 184, 238, 282, 283, 284, 307, 308
Kerr, Sister Agnes 44, 57
King, Olive 35, 44, 48
Kurdistan 244-52

L

Laidlaw (née Courtney), Noelle 169-70
Land Headquarters Medical Research Institute 95, 97, 99
Langford, Major (later Lieutenant Colonel) Caitlin 268-9
Langlois, Major Susie Le P. 229-36, 259
Little, Dr Elaine 63-4

M

MacKenzie, Major Winifred 80-5, 111, 149
Mackerras, Major Josephine 93-101, 150, 154, 306, 308
Malaria 57-8, 95-100, 124-5, 129, 145, 150, 229, 249
Malpass, Captain Alison 267-71
Manunda 119-22
Maribyrnong, munitions research 103-4
Marks, Lieutenant (later Captain) Joanne 236-43
Matthews, Corporal (later Warrant Officer) Elizabeth 183-5
Maughan (-Sutherland), Barbara 170
McArthur Campbell, Alison 113-114, 116, 123-4
MEAO (Middle East Area of Operations) 271, 253, 265
MG (Medal for Gallantry) 209, 215, 314
MINURSO (UN Mission for the Referendum in Western Sahara) 190-8
MM (Military Medal) 59, 62-2, 68, 312

Index

Morris, Captain Jane 245
Murray, Dr Flora 21, 24, 70

N
Neuhaus, Susan, *see* Evans
Nightingale, Florence 2-3, 13
Norris, Frank 26, 28-9
NSC (Nursing Service Cross) 202
NUWSS (National Union of Women's Suffrage Societies) 33-4

O
Occupational Therapy 110, 134-5
Operation Provide Comfort 250
O'Reilly, Dr Susie 14

P
Pearn, Professor (Major General) John x-xii, 302
Physiotherapy 65-7, 110-24, 157-67, 227, 265-7, 297-9
PMG (Peace Monitoring Group) 218-9, 223, 224
PST (Parachute Surgical Team) 282, 285-7, 294

Q
QMAAC, Queen Mary's Army Auxiliary Corps 60-2

R
Rae, Lieutenant Shirley 167
RAMSI (Regional Assistance Mission Solomon Islands) 240, 243
Refshauge, Captain Joan 143-147, 241
Richardson, Lieutenant Gwenyth 127, 129
RMC (Royal Military College, Duntroon) 238
Royaumount 12, 35-44, 49, 71
Rwanda 173, 203-216, 227, 254

S
St Sava, Order of 56-58
Scantlebury (-Brown), Dr (Captain) Vera 20, 25-30, 70, 306
Scott, Dr Jessie (NZ) 44, 48, 57
Serbia 19, 34-38, 44-58, 65-72, 302, 307
Sexton, Dr (Majeure) Helene 13, 32, 33, 46
Skewes(-Fairhead), Lieutenant Di 151, 157-159, 165, 266
Smith, Lieutenant Colonel Fran 278, 279
Solomon Islands 173, 202, 225, 237-243, 254, 287
Somalia 173, 198, 199-206
Somerville, Lieutenant Joan 118-122
Somme 36, 40
(SWH) Scottish Women's Hospitals 19, 33-48, 53, 57, 65, 168
Sumatra Assist 293, 295
Summerscales, Corporal (later Captain) Kerry 202, 221-225, 241, 242

T

Tarin Kowt 266, 273-379
Third Serbian Army 50, 54, 56
Thornton (née Kent-Hughes), Major Mary 84-92, 241, 308
Tran, Captain Tam 244-250, 309
Transfusion 101-106, 126, 178, 221, 223
Trotula 6, 7
Turner(-Cats), Lieutenant Caroline 130-134
Typhoid 47, 56, 249

U

UNMILHOSP (United Nations Military Hospital) 226, 227, 230, 233, 239
UNTAC (United Nations Transitional Authority in Cambodia) 174-188, 308

V

VAD (Voluntary Aid Detachments) 16, 85, 106, 132, 135-137, 140, 180
VD (venereal disease) 71, 144, 145
Val-de-Grâce 32
Van der Rijt, Major Carmel 186, 187
Vaughan-Evans, Captain Carol 204, 209-216, 314
Victoria, Queen 1, 10, 11
Vietnam war 152, 164, 168, 170, 172, 212, 215, 228, 247, 254, 303, 304, 368
Villiers-Cotterêts 37, 43, 44
Vrancic, Flight Lieutenant Cindy 219, 220

W

Walter and Eliza Hall Institute 93, 101, 105
Walker, Dr Mary 8-10
Western Front 12, 16, 20, 29, 30, 35, 41, 47, 48, 64, 68, 137, 138, 227, 302
Western Sahara 173, 188-197, 303
Whelan, Brigadier Georgeina 281, 290, 291, 293, 294, 295, 298, 299, 308, 310, 316
Wilson, Honor 67, 109-112, 159
Wimereux 23
Winter, Lieutenant Colonel Su 252, 255-258, 278, 309
Woolley, Lieutenant Susan 166-167
Wunch, Sergeant (later Lieutenant) Elizabeth 129-131
Wunderly, Captain Alice 131

X

X-rays 22, 43, 49, 82, 85, 90, 91, 169, 218, 221, 231, 234, 238, 239, 269, 286

About the authors

Maggie Elliott Photography

Susan Neuhaus is a surgeon. She has completed a 20-year military career in both the Australian Regular Army and the Army Reserve as a clinician and a commander, serving in Cambodia, Bougainville and Afghanistan. In 2009, she was awarded the Conspicuous Service Cross. She is Clinical Associate Professor, University of Adelaide.

Sharon Mascall-Dare is a journalist. As a BBC radio documentary producer, she has won a number of awards in Australia and overseas. She is currently serving as a Military Public Affairs Officer in the Australian Army Reserve and is Adjunct Associate Professor of Journalism, University of Canberra.

www.ingramcontent.com/pod-product-compliance
Lightning Source LLC
Chambersburg PA
CBHW062030290426
44109CB00026B/2587